Lecture Notes in Mathematics

Edited by A. Dold and B. Eckmann

623

Ivan Erdelyi
Ridgley Lange

Spectral Decompositions on Banach Spaces

Springer-Verlag
Berlin Heidelberg New York 1977

Authors
Ivan Erdelyi
Department of Mathematics
Temple University
Philadelphia, PA 19122/USA

Ridgley Lange
Department of Mathematics
University of New Orleans
New Orleans, LA 70122/USA

AMS Subject Classifications (1970): 47A10, 47A15, 47A60, 47A65, 47B99

ISBN 3-540-08525-4 Springer-Verlag Berlin Heidelberg New York
ISBN 0-387-08525-4 Springer-Verlag New York Heidelberg Berlin

Printed in Germany

Printing and binding: Beltz Offsetdruck, Hemsbach/Bergstr.
2140/3140-543210

FOREWORD

There is a new trend developing in the spectral theory of linear operators. In contrast to the classical spectral theory of linear operators and to the Dunford-type spectral operators which depend on some algebraic and topological structures outside their domains of definition, the contemporary spectral decomposition is defined only in regard to the operators invariant subspaces. In this way, the spectral theory can be conceived as an axiomatic system functioning within the underlying Banach space with possible extensions to more general topological vector spaces.

Our purpose in this work is to extend and unify the intrinsic axiomatic perspective on spectral decompositions. In such extension we wish to consider the widest feasible generalization of the notion "spectral decomposition" in order to learn more about the special cases. In this spirit we start in Chapter II the study of the most abstract form of spectral decomposition so that when we come to the more special theory of "decomposable operators" (Chapter IV) we find that many of the known results of the latter theory are easy consequences of the preceding material. More importantly, however, we obtain solutions to deep problems which have been open and vigorously studied (e.g. the dual theory).

Chapter I presents various classes of invariant subspaces a given operator may have. Special attention is devoted to the single-valued extension property as an essential tool in the study of spectral decompositions. Chapter II is the foundation of our axiomatic attack in the general problem of spectral decomposition. We show that the single-valued extension property is an intrinsic element of the spectral decomposition. The more recent theories of "asymptotic spectral decompositions" are treated in Chapter III. Chapter IV brings the full power of the spectral decomposition to bear in the theory of duality.

The Appendix is aimed to supplement a few topics in the context of spectral decompositions. The example given in section A.1 offers the opportunity to develop some spectral features of the multiplication operator in an elementary way. Section A.2 provides an additional tool, the set-spectrum, for some proving techniques. Section A.3 gives a link between the two major properties which are present in every aspect of the spectral decomposition problem, namely the single-valued extension property and the approximate point spectrum. Finally, section A.4 lists some open problems of the theory to open the way to further exploration.

The main prerequisite for reading these Lecture Notes is the reader's interest in the spectral decomposition problem. The reader interested in this topic will be familiar with some classical properties of linear operators as the open mapping, the closed graph and the Hahn-Banach theorems. Except for a few other theorems known for at least twenty years with references given from Dunford and Schwartz's "Linear Operators" the material presented herein is self-contained.

As a final word, we wish to express our appreciation to Mrs. Geraldine S. Ballard for the neat and careful typing of the manuscript.

The authors.

NOTATIONS

$\mathbb{C}$,	the complex field (plane).
	For a set U:
U^0,	the interior
$\overline{U}$,	the closure
U^c,	the complement (in a given total set)
∂U,	the boundary
$d(\lambda,U)$	the distance from a point λ to U.
	For a linear operator T on a Banach space X:
D_T,	the domain
Ker T,	the kernel (null manifold)
T^*,	the dual (conjugate) operator on X^*
$\tilde{T} = T^{**}$,	the second dual on $\tilde{X} = X^{**}$
$\sigma(T)$,	the spectrum
$\sigma_a(T)$,	the approximate point spectrum
$\sigma_p(T)$,	the point spectrum
$\sigma(x,T)$,	the local spectrum
$\rho(T)$,	the resolvent set
$\rho(x,T)$,	the local resolvent set
$R(\cdot;T)$,	the resolvent operator
$\tilde{x}_T$,	($\tilde{x}$ if T is understood), the maximal analytic extension of $R(\cdot;T)x$, $x \in X$
$X_T(H)$,	definition p. 5
Inv(T),	the family of invariant subspaces under T
Inv(T,F),	definition p. 26
AI(T),	the family of analytically invariant subspaces under T, Definition 2.7, p. 16
SM(T),	the family of spectral maximal spaces of T, Definition 3.1, p. 26.
	For a subspace (closed linear manifold) Y and a linear operator T with $D_T \subset Y$:
$T\|Y$,	the restriction of T to Y
T^Y,	the coinduced operator on the quotient space X/Y,
$Y^\perp$,	the annihilator of Y
$\hat{x}$,	a coset (vector) of the quotient space X/Y.

Other notations:

I,	the identity operator
B(X),	the Banach algebra of bounded linear operators defined on a Banach space X
D(X),	the class of decomposable operators defined on a Banach space X, Definition 11.1, p. 73.
$\mathcal{F}$,	the family of closed subsets of $\mathbb{C}$
$\mathcal{K}$,	the family of compact subsets of $\mathbb{C}$
$\mathcal{S}$(X),	the family of subspaces of a Banach space X
A_T,	the algebra of analytic functions on an open neighborhood of $\sigma(T)$, $T \in B(X)$
$A_T(K)$,	the algebra of analytic functions on a compact K valued in the subclass of operators from B(X) which commute with $T \in B(X)$
C[a,b],	the algebra of continuous complex-valued functions defined on a closed interval [a,b]
supp f,	the support of $f \in C[a,b]$
$\mathcal{E}$,	spectral capacity, Definition 8.1, p. 60
supp $\mathcal{E}$,	support of $\mathcal{E}$, p. 61.

Abbreviations:

λ-T,	for $\lambda I-T$
λ-T|Y,	for $\lambda I\|Y - T\|Y$
SDP (2-SDP),	(2-) spectral decomposition property
SVEP,	the single-valued extension property
det,	determinant
$\bigvee$,	span
c.l.m.	the smallest closed linear manifold spanned by a family of vectors
$\square$	end of proof.

The arrow has two uses:

(a) $x_n \to x$ indicates that the sequence x_n tends to the limit x (in a given topology),

(b) it expresses a mapping between two sets which can be a linear operator, a function, a relation between families of sets, etc.

Except for the standard notations of real intervals and references, the round and square brackets are indiscriminately used for the convenience of a clearer separation of various terms.

CONTENTS

INTRODUCTION

"Concerning general non-normal transformations, it is quite easy to describe the state of our knowledge; it is non-existent. No even unsatisfactory generalization exists for the triangular form or for the Jordan canonical form ..."

P.R. Halmos

from "Finite-Dimensional Vector Spaces" D. Van Nostrand Co., Princeton, 2nd Edition 1958, Appendix, p.192.

The spectral theory of linear operators on some organized topological vector space has undergone a prodigious development from the time Halmos wrote the comments quoted above from his popular book.

A unified treatment for various classes of linear operators which perform a spectral decomposition of the underlying space and give rise to some functional calculi, impels for an axiomatic formulation of the problem. We shall confine the presentation of this problem to bounded and some closed linear operators acting on an abstract Banach space X.

The basic requirement imposed on operators by any spectral theory is the existence of proper invariant subspaces. A proper invariant subspace Y under a given linear operator T may be an element of the spectral decomposition. Also Y produces the restriction $T|Y$ of T as well as the coinduced operator T^Y on the quotient space X/Y, operators which may inherit some basic properties of T.

For a linear operator T, the functional calculus is based on the isomorphic mapping $f \to f(T)$ from the algebra A_T of analytic functions defined on an open neighborhood of $\sigma(T)$ into the Banach algebra B(X), for which $1 \to I$ and $\lambda \to T$, where 1 and λ denote the constant function $f(\lambda) \equiv 1$ and the identity function $f(\lambda) \equiv \lambda$, respectively. An indication that an intimate relation exists between T and f(T) is given by the spectral mapping theorem which asserts that

$$\sigma(f(T)) = f(\sigma(T)).$$

It is known that for every compact $K \subset C$ and open $D \subset K$ there exists an open set G such that

(i) $K \subset G \subset \bar{G} \subset D$;

(ii) G has at most a finite number of components $\{G_i\}_1^n$;

(iii) every component G_i has a boundary formed by a finite number of simple rectifiable Jordan curves Γ_{ij};

(iv) $K \cap \Gamma_{ij} = \emptyset$, for all i,j.

We denote by

$$\Gamma = \bigcup_{i,j} \Gamma_{ij}$$

and for each $f \in A_T$, we put

$$\int_\Gamma f(\lambda)d\lambda = \sum_{i,j} \int_{\Gamma_{ij}} f(\lambda)d\lambda \ .$$

We call Γ, endowed with the above properties, an *admissible contour* which surrounds K and is contained in D.

For $T \in B(X)$ and $K = \sigma(T)$, Dunford's formula

$$f(T) = \frac{1}{2\pi i} \int_\Gamma f(\lambda)R(\lambda;T)d\lambda$$

establishes the isomorphic mapping $f \to f(T)$ of the functional calculus.

If $\sigma(T)$ is disconnected then

$$E = \frac{1}{2\pi i} \int_\Gamma R(\lambda;T)d\lambda$$

is a projection whenever the admissible contour Γ disconnects the spectrum. The range EX of E is a subspace of X invariant under both T and f(T).

CHAPTER I

INVARIANT SUBSPACES

§ 1. *Invariant subspaces and the single-valued extension property.*

The concept of *single-valued extension property* is a major unifying theme for a wide variety of linear operators in the spectral decomposition problem.

1.1. *Definition. A closed linear operator* $T : D_T \ (\subset X) \to X$ *is said to have the single-valued extension property* (abbrev. SVEP) *if for every function* $f : D_f \ (\subset \mathbb{C}) \to D_T$ *analytic on* D_f, *the condition*

$$(\lambda - T)f(\lambda) = 0 \text{ on } D_f$$

implies $f = 0$.

Equivalently, for every $x \in D_T$ any two analytic extensions f,g of $R(\lambda;T)x$ agree on $D_f \cap D_g$. When this property holds, the union of the sets D_f as f varies over all analytic extensions of $R(\lambda;T)x$ is called the local resolvent set and is denoted by $\rho(x,T)$. The SVEP implies the existence of a maximal analytic extension $\tilde{x}(\cdot)$ of $R(\cdot;T)x$ to $\rho(x,T)$. This function identically verifies the equation

$$(\lambda - T)\tilde{x}(\lambda) = x \text{ on } \rho(x,T). \tag{1.1}$$

The local spectrum $\sigma(x,T)$, defined as the complement in $\mathbb{C}$ of $\rho(x,T)$ is the set of the singularities of $\tilde{x}$.

Not every bounded or unbounded linear operator enjoys the SVEP as it is seen from the following useful property due to Finch [1].

1.2. *Proposition. A closed linear operator* T *which is surjective but not injective does not have the* SVEP.

Proof. Given T as stated by the Proposition we can exhibit a function f that violates Definition 1.1. Since T is not injective we can choose $x_0 \in \operatorname{Ker} T$ with $\| x_0 \| = 1$. T being closed and surjective, by the open mapping theorem, there exists $k > 0$ such that for every $y \in X$ there is an $x \in D_T$ satisfying conditions

$$Tx = y, \ \| x \| \le k \| y \| .$$

Choose x_n inductively so that

$$Tx_n = x_{n-1}, \ \| x_n \| \le k \| x_{n-1} \| , \ n = 1,2,\ldots$$

Define

$$(1.2) \qquad f(\lambda) = \sum_{n=0}^{\infty} \lambda^n x_n .$$

Since $\| x_n \| \le k^n$, the series (1.2) converges for $|\lambda| < k^{-1}$ and then f is analytic on

$$D = \{\lambda : |\lambda| < k^{-1}\}.$$

We have

$$(\lambda - T) \sum_{n=0}^{N} \lambda^n x_n = \lambda^{N+1} x_N$$

and

$$\| \lambda^{N+1} x_N \| \le k^N |\lambda|^{N+1} \to 0 \quad \text{when } N \to \infty.$$

Since λ-T is closed,

$$f(\lambda) = \lim_{N \to \infty} \sum_{n=0}^{N} \lambda^n x_n$$

is in the domain of λ-T and

$$(\lambda - T) f(\lambda) = 0 \text{ on } D.$$

Obviously f, as defined by (1.2) is not the zero function. □

1.3. *Corollary. If the closed linear operator* T *has the* SVEP *then* $\lambda \in \sigma(T)$ *iff* λ-T *is not surjective.*

An interesting effect of the SVEP on the dual operator T^* regards the structure of the spectrum.

1.4. *Corollary. If the closed, densely defined linear operator* T *has the* SVEP *then*

$$\sigma_a(T^*) = \sigma(T^*).$$

Similarly, if T^* *has the* SVEP *then*

$$(1.3) \qquad \sigma_a(T) = \sigma(T).$$

Proof. Given T as stated by the Corollary, let $\lambda \in \sigma(T)$. Then λ-T is not surjective. Suppose that λ-T^* is bounded below, i.e. there exists $k > 0$ such that

$$\| (\lambda - T^*) x^* \| \ge k \| x^* \| , \text{ for all } x^* \in X^* .$$

The range of λ -T^* being closed, λ-T^* is injective. But then the kernel of λ-T^* is the zero point and consequently the range of λ-T is the entire space X. This, however, contradicts the hypothesis. Thus λ-T^* is not bounded below and hence $\lambda \in \sigma_a(T^*)$. The second part of the Corollary follows by a similar proof. □

Relation (1.3) will be shown (Theorem 4.5) to be an intrinsic property of operators which have a general spectral decomposition.

Some immediate implications of the SVEP are expressed by the following

1.5. Proposition. If $T \in B(X)$ *has the* SVEP *then the following assertions hold:*

(i) $\sigma(x+y,T) \subset \sigma(x,T) \cup \sigma(y,T)$, $x,y \in X$;

(ii) $a\tilde{x}(\lambda) + b\tilde{y}(\lambda) = (\widetilde{ax+by})(\lambda)$, $a,b \in C$, $x,y \in X$, $\lambda \in \rho(x,T) \cap \rho(y,T)$;

(iii) $\sigma(x,T) = \emptyset$ *iff* $x = 0$;

(iv) $\sigma(Sx,T) \subset \sigma(x,T)$, *for every* $S \in B(X)$ *which commutes with* T;

(v) $\sigma[\tilde{x}(\lambda),T] = \sigma(x,T)$, $x \in X$, $\lambda \in \rho(x,T)$.

Proof. Since properties (i) - (iv) are well-known (see e.g. Dunford and Schwartz [1, XVI., 2.1 and 2.2]) we shall prove (v).

For every $\lambda \in \rho(x,T)$ there is an X-valued function $\tilde{\xi}_\lambda$ analytic verifying equation

$$(\mu-T)\tilde{\xi}_\lambda(\mu) = \tilde{x}(\lambda) \quad \text{on} \quad \rho[\tilde{x}(\lambda),T]. \tag{1.4}$$

Apply $(\lambda-T)$ to both sides of (1.4), use (1.1)

$$(\mu-T)(\lambda-T)\tilde{\xi}_\lambda(\mu) = (\lambda-T)\tilde{x}(\lambda) = x,$$

note that $(\lambda-T)\tilde{\xi}_\lambda$ is analytic on $\rho[\tilde{x}(\lambda),T]$ and conclude that $\mu \in \rho(x,T)$. Thus

$$\sigma(x,T) \subset \sigma[(\tilde{x}(\lambda),T].$$

To obtain the opposite inclusion define the analytic function $g_\lambda:\rho(x,T) \to X$ by

$$g_\lambda(\mu) = \begin{cases} -\dfrac{\tilde{x}(\mu)-\tilde{x}(\lambda)}{\mu-\lambda}, & \text{if } \mu \neq \lambda, \\ -\tilde{x}'(\lambda), & \text{if } \mu = \lambda. \end{cases} \tag{1.5}$$

For $\mu \neq \lambda$ we obtain

$$(\mu-T)g_\lambda(\mu) = -\frac{x}{\mu-\lambda} + \tilde{x}(\lambda) + \frac{x}{\mu-\lambda} = \tilde{x}(\lambda),$$

and this extends by $\mu \to \lambda$. Consequently,

$$\sigma[\tilde{x}(\lambda),T] \subset \sigma(x,T). \quad \square$$

Given $T \in B(X)$, for every set $H \subset C$,

$$X_T(H) = \{x \in X : \sigma(x,T) \subset H\} \tag{1.6}$$

is a linear manifold in X (Proposition 1.5, (i) and (ii)). For K compact in C, denote by $A_T(K)$ the set of functions analytic on a neighborhood of K and valued in the class of operators from B(X) which commute with T. $A_T(K)$ can be extended to an algebra as the one mentioned in the Introduction. For every $f \in A_T(K)$, consider the mapping

$$f:X_T(K) \to X$$

defined by

$$f[T|X_T(K)] = \frac{1}{2\pi i} \int_\Gamma f(\lambda)\tilde{x}(\lambda)d\lambda \,, \tag{1.7}$$

where Γ is an admissible contour surrounding $\sigma(x,T)$ and contained in $\rho(x,T)$.

A functional calculus which parallels the Riesz-Dunford functional calculus can be developed in terms of the local resolvent $\tilde{x}$ by (1.7). Actually, that mapping is homomorphic with $\lambda \to T|X_T(K)$ and $1 \to I|X_T(K)$, (Apostol [6]). This functional calculus gives rise to certain "localization" theorems extensively developed by Bartle [1,2], Bartle and Kariotis [1] and applied to some special operators by Stampfli [1,2]. In the next theorem which generalizes the spectral mapping theorem, we follow the proof of Bartle and Kariotis [1].

1.6. Theorem. Given $T \in B(X)$, *let* $f:D \to C$ *be analytic on an open neighborhood* D *of* $\sigma(T)$. *If both* T *and* $f(T)$ *have the* SVEP *then for every* $x \in X$,

$$f[\sigma(x,T)] = \sigma[x,f(T)]. \tag{1.8}$$

Proof. If f is constant (1.8) is trivially satisfied, assume therefore that f is nonconstant. First, we show that $f(\lambda_0) \in \rho[x,f(T)]$ implies that $\lambda_0 \in \rho(x,T)$. Let $\lambda_0 \in D$ with $f(\lambda_0) \in \rho[x,f(T)]$ and let G be an open neighborhood of λ_0 such that $f(G) \subset \rho[x,f(T)]$. Then

$$[f(\lambda)-f(T)]\tilde{x}[f(\lambda)] = x, \text{ for all } \lambda \in G. \tag{1.9}$$

Since f is nonconstant on G, the function $g:G \times G \to C$ defined by

$$g_\lambda(\mu) = \begin{cases} \dfrac{f(\lambda)-f(\mu)}{\lambda - \mu}, & \text{for } \mu \neq \lambda \\ f'(\lambda), & \text{for } \mu = \lambda \end{cases} \tag{1.10}$$

is analytic in both λ and μ. Note the analogy of the functions defined by (1.5) and (1.10). The functional calculus applied to (1.10) produces

$$f(\lambda)-f(T) = (\lambda-T)g_\lambda(T). \tag{1.11}$$

With the help of (1.11), equation (1.9) becomes

$$[f(\lambda)-f(T)]\tilde{x}[f(\lambda)] = (\lambda-T)g_\lambda(T)\tilde{x}[f(\lambda)] = x$$

and since $g_\lambda(T)\tilde{x}[f(\lambda)]$ is analytic on $G(\ni \lambda_0)$, we have $\lambda_0 \in \rho(x,T)$. Then $\lambda_0 \in \sigma(x,T)$ implies that $f(\lambda_0) \in \sigma[x,f(T)]$ and hence

$$f[\sigma(x,T)] \subset \sigma[x,f(T)].$$

In order to obtain the opposite inclusion, let $\nu_0 \notin f[\sigma(x,T)]$. We separate ν_0 from $f[\sigma(x,T)]$ by disjoint neighborhoods V and W of ν_0 and $f[\sigma(x,T)]$, respectively. Let $H \subset D$ be a neighborhood of $\sigma(x,T)$ such that $f(H) \subset W$, and let Γ be an admissible contour surrounding $\sigma(x,T)$ and contained in H. Then $f(\Gamma) \subset W$ and we have $\nu - f(\lambda) \neq 0$ for all $\nu \in V$ and $\lambda \in H$. Denote $C = \{\lambda : |\lambda| = \|T\| + 1\}$. The functional calculus in $A_T[\sigma(x,T)]$, with the help of (1.11) gives

$$[\nu - f(T)]\frac{1}{2\pi i}\int_\Gamma [\nu - f(\lambda)]^{-1}\tilde{x}_T(\lambda)d\lambda =$$

$$= \frac{1}{2\pi i}\int_\Gamma \tilde{x}_T(\lambda)d\lambda + \frac{1}{2\pi i}\int_\Gamma [f(\lambda)-f(T)][\nu - f(\lambda)]^{-1}\tilde{x}_T(\lambda)d\lambda =$$

$$= \frac{1}{2\pi i}\int_C R(\lambda;T)xd\lambda + \frac{1}{2\pi i}\int_\Gamma [\nu - f(\lambda)]^{-1}g_\lambda(T)xd\lambda = x,$$

the last integral being analytic in the region bounded by Γ. Since

$\frac{1}{2\pi i}\int_\Gamma [\nu - f(\lambda)]^{-1}\tilde{x}_T(\lambda)d\lambda$ is analytic on the complement of $f[\sigma(x,T)]$, it follows that $\nu_0 \in \rho[x,f(T)]$. Thus we have obtained

$$\sigma[x,f(T)] \subset f[\sigma(x,T]. \quad \square$$

1.7. Corollary. Given $T \in B(X)$, *let* $f : D \to \mathbb{C}$ *be analytic on an open neighborhood* D *of* $\sigma(T)$ *and nonconstant on every component of* D. *If both* T *and* $f(T)$ *have the* SVEP *then for every* $F \in \mathcal{F}$,

$$X_{f(T)}(F) = X_T[f^{-1}(F)]. \tag{1.12}$$

Proof. By Theorem 1.6, for every $x \in X_{f(T)}(F)$,

$$f[\sigma(x,T)] = \sigma[x,f(T)] \subset F, \quad F \in \mathcal{F}$$

and consequently $\sigma(x,T) \subset f^{-1}(F)$. Thus $x \in X_T[f^{-1}(F)]$ and hence

$$X_{f(T)}(F) \subset X_T[f^{-1}(F)].$$

Conversely, for $x \in X_T[f^{-1}(F)]$, we have $\sigma(x,T) \subset f^{-1}(F)$ and then Theorem 1.6 implies

$$\sigma[x,f(T)] = f[\sigma(x,T)] \subset f[f^{-1}(F)] = F.$$

Thus

$$X_T[f^{-1}(F)] \subset X_{f(T)}(F),$$

and property (1.12) follows. $\square$

We shall see later (Corollary 2.21) that in Corollary 1.7, the hypotheses that both T and f(T) have the SVEP are redundant.

In the search for invariant subspaces the SVEP may be very helpful. Actually, it provides us with *hyperinvariant* subspaces, i.e. with subspaces that are invariant under every operator which commutes with the given one.

1.8. Proposition. Let $T \in B(X)$ *have the* SVEP. *For every subset* H *of* C, *the subspaces* $\overline{X_T(H)}$ *and* $X_T(H)^{\perp}$ *are hyperinvariant under* T *and* T^*, *respectively.*

Proof. First let $x \in \overline{X_T(H)}$. There is a sequence $\{x_n\} \subset X_T(H)$ which converges (in the norm topology) to x. If $S \in B(X)$ commutes with T then Proposition 1.5 (iv) implies that

$$\sigma(Sx_n,T) \subset \sigma(x_n,T) \subset H, \text{ for all } n.$$

Thus, for every n,

$$Sx_n \in X_T(H)$$

and the continuity of S implies that $Sx \in \overline{X_T(H)}$. Hence, $S\overline{X_T(H)} \subset \overline{X_T(H)}$.

Next, let $y^* \in X_T(H)^{\perp}$. By the first part of the proof, for every $x \in \overline{X_T(H)}$

$$< Sx,y^* > = 0.$$

Consequently, we have

$$0 = < Sx,y^* > = < x,S^*y^* >, \text{ for all } x \in \overline{X_T(H)}$$

and hence $S^*y^* \in X_T(H)^{\perp}$. The proof is complete. □

The relationship between the spectrum and the local spectra is expressed by the following

1.9. Theorem. If $T \in B(X)$ *has the* SVEP *then*

$$\sigma(T) = \bigcup_{x \in X} \sigma(x,T).$$

Proof. The inclusion

$$\sigma(T) \supset \bigcup_{x \in X} \sigma(x,T)$$

follows directly from the definition of the local spectrum. Let

$$\lambda_0 \in \sigma(T) - \bigcup_{x \in X} \sigma(x,T). \tag{1.13}$$

The operator λ_0-T is surjective because for every $x \in X$, we have

$$(\lambda_0-T)\tilde{x}(\lambda_0) = x.$$

Then Corollary 1.3 implies that $\lambda_0 \in \rho(T)$ but this contradicts (1.13). □

The SVEP is inherited by the restrictions of the given operator.

1.10. Proposition. Let $T \in B(X)$ *have the* SVEP *and let* $Y \in \mathrm{Inv}(T)$. *Then* $T|Y$ *has the* SVEP *and*

$$\sigma(y,T) \subset \sigma(y,T|Y), \text{ for every } y \in Y.$$

Proof. The first assertion of the Proposition follows at once. Let $y \in Y$. For every $\lambda \in \rho(y,T|Y)$, we have

$$(\lambda-T)\tilde{y}(\lambda) = (\lambda-T|Y)\tilde{y}(\lambda) = y$$

and hence

$$\rho(y,T|Y) \subset \rho(y,T). \ \square$$

The SVEP is stable under uniform convergence.

1.11. Theorem. Given $T \in B(X)$, *let* $\{T_n\}$ *be a sequence in* $B(X)$ *satisfying*

(i) Each T_n *commutes with* T;

(ii) Each T_n *has the* SVEP;

(iii) The sequence $\{T_n\}$ *converges to* T *in the uniform operator topology.*

Then T *has the* SVEP.

Proof. Let $f:D \to X$ be analytic and verify equation

$$(\lambda-T)f(\lambda) = 0 \text{ on } D. \tag{1.14}$$

Let $\lambda \in D$ be arbitrary and let

$$K_i = \{\nu \in C: |\nu-\lambda| \le r_i\} \subset D, \ i = 1,2 \text{ and } r_2 < r_1.$$

By the uniform convergence, for every $\delta > 0$ there is an n such that the operator $Q_n = T_n-T$ has the norm

$$\| Q_n \| < \delta.$$

Take $\delta = \min(r_2, r_1-r_2)$ and denote

$$K_\delta = \{\nu: |\nu-\lambda| \le \delta\}.$$

For $\mu \in K_\delta^c$, $\mu-\lambda \in \rho(Q_n)$ and (1.14) becomes successively:

$$(\mu-T_n)f(\lambda) = (\mu-\lambda-Q_n)f(\lambda),$$

(1.15) $$(\mu-T_n)R(\mu-\lambda;Q_n)f(\lambda) = f(\lambda).$$

Since $\mu \to R(\mu-\lambda;Q_n)f(\lambda)$ is analytic on K_δ^c, it follows that

(1.16) $$\sigma[f(\lambda),T_n] \subset K_\delta.$$

In view of (1.15), by integration along the boundary of K_1, we obtain

$$(\mu-T_n)\frac{1}{2\pi i}\int_{\partial K_1}\frac{R(\mu-\nu;Q_n)f(\nu)}{\nu-\lambda}\,d\nu = \frac{1}{2\pi i}\int_{\partial K_1}\frac{f(\nu)}{\nu-\lambda}\,d\nu = f(\lambda).$$

The function

$$\mu \to \frac{1}{2\pi i}\int_{\partial K_1}\frac{R(\mu-\nu;Q_n)f(\nu)}{\nu-\lambda}\,d\nu$$

is clearly analytic on $\overset{0}{K}_2$, and therefore

(1.17) $$\sigma[f(\lambda),T_n] \subset (\overset{0}{K}_2)^c \subset K_\delta^c .$$

It follows from (1.16) and (1.17) that

$$\sigma[f(\lambda),T_n] = \emptyset$$

and then Proposition 1.5 (iii) implies that $f(\lambda) = 0$. Since λ is arbitrary in D, we conclude that $f = 0$. □

The property expressed by the foregoing theorem holds under weaker conditions. Condition (i) can be skipped under a slightly different topology.

1.12. Corollary. If $T \in B(X)$ *has the* SVEP *and* Q *is quasinilpotent commuting with* T *then* $T+Q$ *has the* SVEP.

Proof. Let $f:D \to X$ be analytic on an open $D \subset \mathbb{C}$ and verify equation

$$(\lambda-T-Q)f(\lambda) = 0 \text{ on } D.$$

For $\mu \neq \lambda$, write

$$(\mu-T)f(\lambda) = (\mu-\lambda+Q)f(\lambda)$$

and then follow the proof of Theorem 1.11 from (1.15) to the end by interpreting $Q_n = -Q$, $T_n = T$ and $K_\delta = \{\lambda\}$. □

The SVEP is stable under finite direct sums.

1.13. Theorem. Let $T_i \in B(X_i)$, $i = 1,2$. $T_1 \oplus T_2$ *has the* SVEP *iff both* T_1 *and* T_2 *have that property. Moreover,*

$$\sigma(x_1 \oplus x_2, T_1 \oplus T_2) = \sigma(x_1,T_1) \cup \sigma(x_2,T_2).$$

Proof. First, assume that T_1 and T_2 have the SVEP and let $f = f_1 \oplus f_2 : D \to X_1 \oplus X_2$ be analytic on an open $D \subset \mathbb{C}$, with $f_i : D \to X_i$, $(i = 1,2)$ analytic on D. The condition

$$[\lambda-(T_1 \oplus T_2)]f(\lambda) = 0 \text{ on } D$$

implies

$$(\lambda-T_i)f_i(\lambda) = 0 \text{ on } D,\ i = 1,2.$$

By the SVEP of T_1 and T_2, we have $f_1 = 0$ and $f_2 = 0$. Thus $f = 0$.

Next, assume that $T_1 \oplus T_2$ has the SVEP and let $f_i : D \to X_i$ be analytic and verify equations

$$(\lambda-T_i)f_i(\lambda) = 0 \text{ on } D,\ i = 1,2.$$

Then

$$0 = (\lambda-T_1)f_1(\lambda) \oplus (\lambda-T_2)f_2(\lambda) = [\lambda-(T_1 \oplus T_2)][f_1(\lambda) \oplus f_2(\lambda)] \text{ on } D$$

and by the SVEP of $T_1 \oplus T_2$ we obtain

$$f_1(\lambda) \oplus f_2(\lambda) = 0 \text{ on } D.$$

Thus, $f_1 = 0$ and $f_2 = 0$.

Now let $\lambda \in \rho(x_1 \oplus x_2, T_1 \oplus T_2)$. There is a neighborhood D of λ and an analytic function $f = f_1 \oplus f_2 : D \to X_1 \oplus X_2$ (with f_1, f_2 analytic) on D such that

$$(\lambda-T_1)f_1(\lambda) \oplus (\lambda-T_2)f_2(\lambda) = [\lambda-(T_1 \oplus T_2)]f(\lambda) = x_1 \oplus x_2.$$

Then

$$(\lambda-T_i)f_i(\lambda) = x_i,\ i = 1,2$$

and hence $\lambda \in \rho(x_1,T_1) \cap \rho(x_2,T_2)$. Thus, we have

$$\sigma(x_1,T_1) \cup \sigma(x_2,T_2) \subset \sigma(x_1 \oplus x_2, T_1 \oplus T_2).$$

The opposite inclusion has a similar proof. $\square$

Given $T \in B(X)$, a subspace $Y \in \operatorname{Inv}(T)$ produces two related linear operators: the restriction $T|Y$ and the coinduced T^Y, the latter acting on the quotient space X/Y. In general, the three spectra $\sigma(T)$, $\sigma(T|Y)$ and $\sigma(T^Y)$ have the property that the union of any two of them contains the third.

1.14. *Proposition.* *Given* $T \in B(X)$, *for every* $Y \in \mathrm{Inv}(T)$ *we have*

(i) $$\sigma(T) \subset \sigma(T|Y) \cup \sigma(T^Y);$$

(ii) $$\sigma(T|Y) \subset \sigma(T) \cup \sigma(T^Y);$$

(iii) $$\sigma(T^Y) \subset \sigma(T) \cup \sigma(T|Y).$$

Proof. (i): Let $\lambda \in \rho(T|Y) \cap \rho(T^Y)$. The equation

$$(\lambda-T)x = 0$$

produces the following implications:

$$(\lambda-T^Y)\hat{x} = 0 \Rightarrow \hat{x} = 0 \Rightarrow x \in Y \Rightarrow (\lambda-T|Y)x = 0 \Rightarrow x = 0.$$

Hence $\lambda-T$ is injective.

Next, let $x \in X$ be arbitrary. There exists $y \in X$ such that

$$(\lambda-T^Y)\hat{y} = \hat{x}.$$

Then

$$(\lambda-T)y - x \in Y$$

and hence there is a vector $u \in Y$ defined by

$$u = R(\lambda;T|Y)[(\lambda-T)y - x].$$

Furthermore, we obtain

$$(\lambda-T)u = (\lambda-T)R(\lambda;T|Y)[(\lambda-T)y-x] = (\lambda-T)y - x$$

and hence

$$(\lambda-T)(y-u) = x.$$

Thus, $\lambda-T$ is surjective and by the previous argument, bijective. This proves that $\lambda \in \rho(T)$.

(ii): Let $\lambda \in \rho(T) \cap \rho(T^Y)$. It is clear that $\lambda-T|Y$ is injective. If $y \in Y$ is arbitrary then there is an $x \in X$ with

$$y = (\lambda-T)x.$$

Passing to the quotient space X/Y, the hypothesis on λ gives $x \in Y$ and consequently $\lambda \in \rho(T|Y)$.

(iii): Let $\lambda \in \rho(T) \cap \rho(T|Y)$. The equation

$$(\lambda-T^Y)\hat{x} = 0$$

implies that $(\lambda-T)x \in Y$ and hence we have

$$x = R(\lambda;T)(\lambda-T)x = R(\lambda;T|Y)(\lambda-T)x \in Y.$$

Thus $\hat{x} = 0$ and hence $\lambda-T^Y$ is injective.

Next, let $\hat{x} \in X/Y$ be arbitrary. For every $y \in Y$, there is a unique $u \in Y$ such that

$$(\lambda-T|Y)u = (\lambda-T)u = y.$$

Also, there is a unique $v \in X$ which verifies equation

$$(\lambda-T)v = x.$$

Summing up, the last two arguments, there is a unique $\hat{z} \in X/Y$ which verifies equation

$$(\lambda-T^Y)\hat{z} = \hat{x},$$

and this proves that $\lambda-T^Y$ is surjective. The bijectivity of $\lambda-T^Y$ implies that $\lambda \in \rho(T^Y)$ and this concludes the proof. □

The *spectral inclusion property*

$$\sigma(T|Y) \subset \sigma(T), \text{ for } Y \in \mathrm{Inv}(T) \tag{1.18}$$

will play an important role in the spectral decomposition problem. There are some necessary and sufficient conditions for this property to hold.

1.15. Proposition. Given $T \in B(X)$, *for every* $Y \in \mathrm{Inv}(T)$ *the following statements are equivalent:*

(i) $\sigma(T|Y) \subset \sigma(T)$;

(ii) $\sigma(T^Y) \subset \sigma(T)$;

(iii) $R(\lambda;T)Y \subset Y$, $\lambda \in \rho(T)$.

Proof. (i) <=> (ii) follows from properties (ii) and (iii) of Proposition 1.14.

(i) => (iii): Let $y \in Y$. For $\lambda \in \rho(T) \subset \rho(T|Y)$, $\lambda-T|Y$ is surjective and hence there is an $x \in Y$ such that

$$y = (\lambda-T)x.$$

The injectivity of $\lambda-T|Y$ implies (iii).

(iii) => (i): For $\lambda \in \rho(T)$, $\lambda-T|Y$ is injective because otherwise the inclusions

$$\lambda \in \sigma_p(T|Y) \subset \sigma_p(T) \subset \sigma(T)$$

contradict the hypothesis. Thus, for every $y \in Y$ there is a unique $x \in X$ which verifies

$$y = (\lambda-T)x,$$

and hence

$$x = R(\lambda;T)y \in Y.$$

Consequently, $\lambda-T|Y$ is bijective and it follows that $\rho(T) \subset \rho(T|Y)$. □

1.16. Corollary. Given $T \in B(X)$ *with the* SVEP, *for every* $Y \in \mathrm{Inv}(T)$, *the following implications hold:*

(i) $\sigma(y,T) = \sigma(y,T|Y)$, for all $y \in Y \Rightarrow \sigma(T|Y) \subset \sigma(T)$;

(ii) $\sigma(y,T) = \sigma(y,T|Y)$, for all $y \in Y \Leftrightarrow \{\tilde{y}_T(\lambda) : \lambda \in \rho(y,T)\} \subset Y$.

Proof. (i): With the help of Theorem 1.9, we obtain

$$\sigma(T|Y) = \bigcup_{y \in Y} \sigma(y,T|Y) \subset \bigcup_{x \in X} \sigma(x,T) = \sigma(T).$$

(ii): If for all $y \in Y$, we have $\rho(y,T|Y) = \rho(y,T)$ and if $\lambda \in \rho(y,T)$ then

$$\tilde{y}_T(\lambda) = \tilde{y}_{T|Y}(\lambda) \in Y.$$

Conversely, if $\tilde{y}_T(\lambda) \in Y$ for all $\lambda \in \rho(y,T)$ then

$$(\lambda - T|Y)\tilde{y}_T(\lambda) = (\lambda - T)\tilde{y}_T(\lambda) = y,$$

and hence $\rho(y,T) \subset \rho(y,T|Y)$. Now, Proposition 1.10 concludes the proof. □

One of the above implications can be strengthened. In fact, Y being a subspace, by (ii) we obtain

$\sigma(y,T) = \sigma(y,T|Y)$, for all $y \in Y \Rightarrow$ c.l.m. $\{\tilde{y}_T(\lambda) : \lambda \in \rho(y,T)\} \subset Y$.

Given $T \in B(X)$ and $Y \in \mathrm{Inv}(T)$, can any bounded component of $\rho(T)$ properly and simultaneously intersect $\sigma(T|Y)$ and $\rho(T|Y)$? The answer is no.

1.17. Proposition. Given $T \in B(X)$, *let* $Y \in \mathrm{Inv}(T)$. *If* G *is any bounded component of* $\rho(T)$ *then*

$$\textit{either } \sigma(T|Y) \cap G = \emptyset \quad \textit{or } G \subset \sigma(T|Y).$$

Proof. Suppose there is a bounded component G of $\rho(T)$ such that

$$\sigma(T|Y) \cap G \neq \emptyset \text{ and } G \not\subset \sigma(T|Y).$$

Then there is a $\lambda \in G$ such that

$$\lambda \in \partial[\sigma(T|Y)] \subset \sigma_a(T|Y) \subset \sigma_a(T) \subset \sigma(T),$$

but this is a contradiction. □

§ 2. *Analytically invariant subspaces.*

For further meaningful applications we must sacrifice some generality. Most of the invariant subspaces employed in spectral decompositions satisfy the spectral inclusion property (1.18). We shall use the terminology of Bartle and Kariotis [1] in the following

2.1. *Definition.* *Given* $T \in B(X)$, $Y \in \mathrm{Inv}(T)$ *is called a* ν*-space for* T *if*

$$\sigma(T|Y) \subset \sigma(T). \tag{2.1}$$

Some equivalent defining conditions for ν-spaces are given by Proposition 1.15.

2.2. *Proposition.* *If* Y *is a* ν*-space for* $T \in B(X)$ *then*

$$\sigma(T) = \sigma(T|Y) \cup \sigma(T^Y). \tag{2.2}$$

Proof. In view of (2.1), Proposition 1.15 (ii) and Proposition 1.14 (i) imply (2.2). □

2.3. *Theorem.* *Given* $T \in B(X)$, *let* $f:D \to C$ *be a function analytic on an open neighborhood* D *of* $\sigma(T)$. *If* Y *is a* ν*-space for* T *then* Y *is a* ν*-space for* $f(T)$. *Furthermore, we have*

$$f(T)|Y = f(T|Y) \text{ and } f(T)^Y = f(T^Y).$$

Proof. Let Y be a ν-space for T. By Proposition 1.15, Y is invariant under $R(\lambda;T)$ and by the functional calculus Y is invariant under $f(T)$. Therefore,

$$f(T)|Y = f(T|Y),$$

and in view of (2.1), the spectral mapping theorem implies the following inclusions

$$\sigma[f(T)|Y] = \sigma[f(T|Y)] = f[\sigma(T|Y)] \subset f[\sigma(T)] = \sigma[f(T)].$$

Next, for $\hat{x} \in X/Y$, by the continuity of the canonical map $X \to X/Y$ and with the help of Proposition 1.15, we obtain successively:

$$f(T)^Y\hat{x} = \widehat{f(T)x} = \widehat{\frac{1}{2\pi i}\int_\Gamma f(\lambda)R(\lambda;T)x\,d\lambda} = \frac{1}{2\pi i}\int_\Gamma f(\lambda)R(\lambda;T)^Y\hat{x}\,d\lambda =$$

$$= \frac{1}{2\pi i}\int_\Gamma f(\lambda)R(\lambda;T^Y)\hat{x}\,d\lambda = f(T^Y)\hat{x},$$

where Γ is an admissible contour surrounding $\sigma(T)$ and contained in $\rho(T) \subset \rho(T^Y)$. Since $\hat{x}$ is arbitrary in X/Y, we have $f(T)^Y = f(T^Y)$. □

2.4. *Theorem.* *Given* $T \in B(X)$, *let* $f:D \to C$ *be a function injective and analytic on an open neighborhood* D *of* $\sigma(T)$. *If* Y *is a* ν*-space for* $f(T)$ *then* Y *is a* ν*-space for* T.

Proof. Let Y be a ν-space for $f(T)$. By Proposition 1.15, Y is invariant under $R[\lambda;f(T)]$. Apply Dunford's theorem on composite operator-valued functions (Dunford [1], Dunford and Schwartz [1, VII. 3.12]) to the composition $f^{-1}\circ f$. For an admissible contour Γ which surrounds $\sigma(T)$ and is contained in $D \cap \rho(T)$, we have

$$f^{-1}[f(T)] = \frac{1}{2\pi i} \int_\Gamma f^{-1}[f(\lambda)]R(\lambda;T)d\lambda = \frac{1}{2\pi i} \int_\Gamma \lambda\, R(\lambda;T)d\lambda = T.$$

On the other hand, we have

$$f^{-1}[f(T)] = \frac{1}{2\pi i} \int_{f(\Gamma)} f^{-1}(\lambda)R[\lambda;f(T)]d\lambda \ .$$

Combining the above results it is easy to show that Y is invariant under T:

$$TY = f^{-1}[f(T)]Y = \frac{1}{2\pi i} \int_{f(\Gamma)} f^{-1}(\lambda)R[\lambda;f(T)]Yd\lambda \subset Y.$$

Now, we conclude the proof through the following inclusions

$$f[\sigma(T|Y)] = \sigma[f(T|Y)] = \sigma[f(T)|Y] \subset \sigma[f(T)] = f[\sigma(T)],$$

$$\sigma(T|Y) \subset \sigma(T). \ \square$$

2.5. *Proposition. Given* $T \in B(X)$, *if* $\sigma(T)$ *does not separate the plane then every invariant subspace is a* ν*-space for* T.

Proof. Any $Y \in Inv(T)$ is invariant under $R(\lambda;T)$ for $|\lambda| > \|T\|$. Thus, for $y \in Y$ and $|\lambda| > \|T\|$, we have $R(\lambda;T)y \in Y$. Since the hypothesis implies that $\rho(T)$ is simply connected, it follows by analytic continuation that $R(\lambda;T)y \in Y$ on all of $\rho(T)$. Thus Y is invariant under $R(\lambda;T)$ for all $\lambda \in \rho(T)$. Then Proposition 1.15 implies the spectral inclusion property (2.1). $\square$

2.6. *Proposition. Every hyperinvariant subspace under* $T \in B(X)$ *is a* ν*-space for* T.

Proof. If Y is hyperinvariant under T then it is invariant under $R(\lambda;T)$ on $\rho(T)$ and then Proposition 1.15 concludes the proof. $\square$

More generally, any subspace invariant under both T and $R(\lambda;T)$ on $\rho(T)$ is a ν-space for T. In particular, if $E \in B(X)$ is a projection commuting with the given T then EX is a ν-space for T.

In order to make the SVEP more useful, we proceed by introducing and studying the first important class of ν-spaces.

2.7. *Definition. Given* $T \in B(X)$, *a subspace* $Y \in Inv(T)$ *is called analytically invariant under* T *if for every function* $f: D \to X$ *analytic on some open* $D \subset C$, *the condition*

$$(\lambda-T)f(\lambda) \in Y \text{ on } D$$

implies that $f(\lambda) \in Y$ *on* D.

We denote by AI(T) the family of analytically invariant subspaces under T.

2.8. *Proposition. Every analytically invariant subspace is a* ν*-space for* $T \in B(X)$.

Proof. Let $Y \in AI(T)$ and let $y \in Y$ be arbitrary. Since

$$y = (\lambda-T)R(\lambda;T)y \text{ on } \rho(T),$$

Definition 2.7 implies that $R(\lambda;T)y \in Y$ on $\rho(T)$ and then Proposition 1.15 concludes the proof. □

2.9. *Proposition. If* $T \in B(X)$ *has the* SVEP *and* $Y \in AI(T)$ *then*

$$\sigma(y,T) = \sigma(y, T|Y), \text{ for all } y \in Y.$$

Proof. Let $y \in Y$ and $\lambda \in \rho(y,T)$. Then

$$(\lambda-T)\tilde{y}(\lambda) = y$$

implies that $\tilde{y}(\lambda) \in Y$ on $\rho(y,T)$ and Corollary 1.16 (ii) concludes the proof. □

2.10. *Corollary. Given* $T \in B(X)$ *with the* SVEP, *let* $Y \in AI(T)$. *Then*

$$\text{c.l.m. } \{\tilde{y}(\lambda) : y \in Y, \lambda \in \rho(y,T)\} \subset Y.$$

The following result gives an important characterization of analytically invariant subspaces and has many key applications in the spectral decomposition theory.

2.11. *Theorem. Given* $T \in B(X)$, *a subspace* $Y \in Inv(T)$ *is analytically invariant under* T *iff the coinduced operator* T^Y *has the* SVEP.

Proof. Assume that T^Y has the SVEP and let $f: D \to X$ be analytic and satisfy condition

$$(\lambda-T)f(\lambda) \in Y \text{ on an open } D \subset \mathbb{C}.$$

By the natural homomorphism, it follows that

$$(\lambda-T^Y)\hat{f}(\lambda) = 0 \text{ on } D$$

and then by the SVEP, $\hat{f} = 0$. Hence $f(\lambda) \in Y$ for all $\lambda \in D$.

Conversely, assume that $Y \in AI(T)$. Let $\hat{f}: D \to X/Y$ be analytic on D and satisfy condition

$$(\lambda-T^Y)\hat{f}(\lambda) = 0 \text{ on } D. \tag{2.3}$$

Without loss of generality we may assume that D is connected. Let

$$\hat{f}(\lambda) = \sum_{n=0}^{\infty} \hat{a}_n(\lambda-\lambda_0)^n, \text{ with } \hat{a}_n \in X/Y$$

be the Taylor series of $\hat{f}$ in a neighborhood of a point $\lambda_0 \in D$. For every n we can choose $a_n \in \hat{a}_n$ such that

$$\| a_n \| \leq \| \hat{a}_n \| + 1.$$

This is possible because in the topology of the quotient space,

$$\| \hat{a} \| = \inf_{a \in \hat{a}} \| a \| .$$

Then

$$\overline{\lim_{n \to \infty}} \| a_n \|^{1/n} \leq \overline{\lim_{n \to \infty}} \| \hat{a}_n \|^{1/n} + 1$$

and hence

$$f(\lambda) = \sum_{n=0}^{\infty} a_n (\lambda-\lambda_0)^n \in \hat{f}(\lambda)$$

is analytic in a neighborhood $D'(\subset D)$ of λ_0. Now (2.3) implies that

$$(\lambda-T)f(\lambda) \in Y \text{ on } D'$$

and since Y is analytically invariant it follows that $f(\lambda) \in Y$. Consequently, $\hat{f}(\lambda) = 0$ on D' and on all of D, by analytic continuation. □

2.12. *Corollary. Given* $T \in B(X)$, *let* Q *be quasinilpotent commuting with* T. *Then every analytically invariant subspace under* T *which is invariant under* Q *is analytically invariant under* T + Q.

Proof. Let $Y \in AI(T)$ be invariant under Q. By Theorem 2.11, T^Y has the SVEP. Q^Y being quasinilpotent and commuting with T^Y, the sum $T^Y + Q^Y = (T + Q)^Y$ has the SVEP by Corollary 1.12. Now by Theorem 2.11, Y is analytically invariant under T + Q. □

2.13. *Corollary. Given* $T \in B(X)$, *let* $\{T_n\}$ *be a sequence of operators in* B(X) *which converges uniformly to* T. *If for every* n, T_n *commutes with* T *and a subspace* Y *is analytically invariant under each* T_n *then* Y *is analytically invariant under* T.

Proof. The fact that Y is invariant under T follows directly from the continuity of the operators. By Theorem 2.11, every T_n^Y has the SVEP. The uniform convergence of T_n to T implies the uniform convergence of T_n^Y to T^Y. Theorem 1.11 applied to the sequence $\{T_n^Y\}$ implies that T^Y has the SVEP and then by Theorem 2.11, $Y \in AI(T)$. □

Some simple examples of analytically invariant subspaces now follow.

2.14. Example. Let $T \in B(X)$ *have the* SVEP. *If* E *is a bounded projection in* X *which commutes with* T *then* EX *is analytically invariant under* T.

Proof. Let $f:D \to X$ be analytic and satisfy condition

$$(\lambda-T)f(\lambda) \in EX \text{ on an open } D \subset \mathbb{C}.$$

Since E is bounded, the function $g:D \to X$ defined by

$$g(\lambda) = (I-E)f(\lambda)$$

is analytic on D. Moreover, since E commutes with T, it follows that

$$(\lambda-T)g(\lambda) = 0 \text{ on } D.$$

By the SVEP, $g(\lambda) = 0$ on D and hence

$$f(\lambda) = Ef(\lambda), \text{ for all } \lambda \in D. \ \square$$

2.15. Example. The kernel of every $T \in B(X)$ *with the* SVEP *is analytically invariant under* T.

Proof. Let $f:D \to X$ be analytic on an open $D \subset \mathbb{C}$ and satisfy condition

$$(\lambda-T)f(\lambda) \in \operatorname{Ker} T, \text{ for all } \lambda \in D.$$

Then

$$0 = T(\lambda-T)f(\lambda) = (\lambda-T)Tf(\lambda) \text{ on } D$$

and by the SVEP,

$$Tf(\lambda) = 0 \text{ on } D. \ \square$$

2.16. Example. Given $T \in B(X)$, let $Y \in AI(T)$ and let $Z = \operatorname{Ker} T$. Then

(i) $\overline{Y + Z}$ is analytically invariant under T;

(ii) $\overline{TY}$ is analytically invariant under T.

Proof. (i): Let $f:D \to X$ be analytic and satisfy condition

$$(\lambda-T)f(\lambda) \in \overline{Y + Z} \text{ on an open } D \subset \mathbb{C}. \tag{2.4}$$

For $\lambda \in D$, there are sequences $\{y_n\}$ and $\{z_n\}$ in Y and Z, respectively such that

$$(\lambda-T)f(\lambda) = \lim_{n \to \infty} (y_n + z_n).$$

Then

$$(\lambda-T)Tf(\lambda) = \lim_{n \to \infty} Ty_n \in Y \text{ on } D.$$

Since $Y \in AI(T)$, $Tf(\lambda) \in Y$ on D. It follows from (2.4) that $\lambda f(\lambda) \in \overline{Y + Z}$, and hence $f(\lambda) \in \overline{Y + Z}$ on D.

(ii): Let $f: D \to X$ be analytic and satisfy condition

$$(\lambda - T)f(\lambda) \in \overline{TY} \text{ on an open } D \subset C.$$

Since $\overline{TY} \subset Y$ and $Y \in AI(T)$, it follows that $f(\lambda) \in Y$ on D. Hence for each $\lambda \in D$,

$$\lambda f(\lambda) = (\lambda - T)f(\lambda) + Tf(\lambda) \in \overline{TY} + TY \subset \overline{TY}. \square$$

The analytically invariant subspaces satisfy certain types of transitivity properties.

2.17. *Proposition.* *Given* $T \in B(X)$, *let* $Y, Z \in Inv(T)$ *with* $Y \subset Z$. *The following properties hold.*

(i) *If* $Y \in AI(T)$ *then* $Y \in AI(T|Z)$; *if* $Y \in AI(T|Z)$ *and* $Z \in AI(T)$ *then* $Y \in AI(T)$.

(ii) *The quotient space* Z/Y *is analytically invariant under* T^Y *iff* Z *is analytically invariant under* T (T^Y denotes the coinduced operator on X/Y).

Proof. (i) is left to the reader.

(ii): Let Z/Y be analytically invariant under T^Y and let $f: D \to X$ be analytic on an open $D \subset C$ and satisfy condition

$$(\lambda - T)f(\lambda) \in Z \text{ on } D.$$

In the quotient space X/Y the map $\lambda \to \hat{f}(\lambda)$ is analytic on D and

$$(\lambda - T^Y)\hat{f}(\lambda) \in Z/Y \text{ on } D.$$

Then, by hypothesis, $\hat{f}(\lambda) \in Z/Y$ and hence $f(\lambda) \in Z$ on D. Thus $Z \in AI(T)$.

Conversely, assume that $Z \in AI(T)$ and let $\hat{f}: D \to X/Y$ be analytic and verify

$$(\lambda - T^Y)\hat{f}(\lambda) \in Z/Y \text{ on } D.$$

We may assume that D is connected. Fix λ_0 in D. By an argument used in the second part of the proof of Theorem 2.11, $\hat{f}$ can be lifted to an X-valued function f analytic on a neighborhood $D' (\subset D)$ of λ_0, i.e. $f(\lambda) \in \hat{f}(\lambda)$ on D'. Then

$$(\lambda - T)f(\lambda) \in Z \text{ on } D',$$

and the hypothesis on Z implies that $f(\lambda) \in Z$ on D'. Passing to the quotient space X/Y, we have $\hat{f}(\lambda) \in Z/Y$ on all of D, by analytic continuation. $\square$

2.18. *Proposition.* *Given* $T_i \in B(X_i)$, *let* $Y_i \in Inv(T_i)$, $i = 1,2$. *The subspace* $Y = Y_1 \oplus Y_2$ *is analytically invariant under* $T = T_1 \oplus T_2$ *iff each* $Y_i \in AI(T_i)$.

Proof. First, assume that each $Y_i \in AI(T)$. In view of Theorem 2.11, each coinduced $(T)^{Y_i}$ on X_i/Y_i has the SVEP. Then Theorem 1.13 implies that $T^Y = (T)^{Y_1} \oplus (T)^{Y_2}$ has the SVEP. Again Theorem 2.11 proves that $Y \in AI(T)$.

Conversely, assume that $Y \in AI(T)$. Each Y_i in the direct sum decomposition of Y is the range of a projection in Y commuting with $T|Y$. By virtue of Example 2.14, each Y_i is analytically invariant under $T|Y$ and hence under T_i by Proposition 2.17 (i). □

Next, we investigate for the stability of analytically invariant subspaces under functional calculus. First we need a lemma.

2.19. *Lemma. Given* $T \in B(X)$, *let* Y *be analytically invariant under* T. *If* $f:D \to X$ *is a nonzero analytic function on an open connected set* D *such that*

$$(\lambda-T)f(\lambda) = 0 \text{ on } D \tag{2.5}$$

then $D \subset \sigma_p(T|Y)$.

Proof. Let G be a nonempty component of $D \cap \rho(T|Y)$ so that there is some $\mu \in G$ with $f(\mu) \neq 0$. Y being analytically invariant, it follows from (2.5) that $f(\mu) \in Y$. Then

$$(\mu-T)f(\mu) = (\mu-T|Y)f(\mu) = 0$$

implies that $f(\mu) = 0$. This contradiction concludes the proof. □

2.20. *Theorem. Given* $T \in B(X)$, *let* $f:G \to C$ *be analytic on an open neighborhood* G *of* $\sigma(T)$. *Then any* $Y \in AI(T)$ *is analytically invariant under* $f(T)$.

Proof. We may assume that G is connected. Let $Y \in AI(T)$. Then Y is invariant under the resolvent and by the functional calculus, Y is invariant under all functions of T which are analytic on some open neighborhood of $\sigma(T)$.

Let $g:D \to X$ be analytic and satisfy condition

$$[\lambda-f(T)]g(\lambda) \in Y \text{ on an open } D \subset C. \tag{2.6}$$

If $D \cap \rho[f(T)] \neq \emptyset$ then the assertion of the Theorem follows at once. Therefore, assume that $D \subset \sigma[f(T)] = f[\sigma(T)]$. For a fixed $\lambda \in D$, the equation

$$\lambda-f(\mu) = 0 \tag{2.7}$$

has at most a finite number of roots in $\sigma(T)$. If we discard the multiple roots (i.e. the zeros of $f'(\mu)$), we have the simple roots $\mu_1,\mu_2,\ldots,\mu_n$ of (2.7) in a disk $D_1 \subset D$. By Rouché's theorem, there is a disk $D_2 \subset D_1$ such that equation (2.7) has the same number of roots $\mu_1(\lambda), \mu_2(\lambda),\ldots, \mu_n(\lambda)$ for every $\lambda \in D_2$.

Note that the functions $\mu_i(\lambda)$, $(1 \leq i \leq n)$ are analytic on D_2. Now we can factor $\lambda - f(\mu)$ as follows:

$$\lambda - f(\mu) = [\mu-\mu_1(\lambda)][\mu-\mu_2(\lambda)]\ldots[\mu-\mu_n(\lambda)]h_\lambda(\mu), \tag{2.8}$$

where h_λ is analytic in μ and nonzero on G for $\lambda \in D_2$. By the functional calculus $h_\lambda(T)$ is invertible in $B(X)$ because

$$0 \notin h_\lambda[\sigma(T)] = \sigma[h_\lambda(T)].$$

The functional calculus applied to (2.8) gives

$$\lambda - f(T) = [T-\mu_1(\lambda)][T-\mu_2(\lambda)]\ldots[T-\mu_n(\lambda)]h_\lambda(T)$$

and then, by virtue of (2.6) we obtain

$$[\lambda - f(T)]g(\lambda) = [T-\mu_1(\lambda)][T-\mu_2(\lambda)]\ldots[T-\mu_n(\lambda)]h_\lambda(T)g(\lambda) \in Y.$$

Y being analytically invariant under T, we obtain

$$h_\lambda(T)g(\lambda) \in Y.$$

Y being invariant under $h_\lambda(T)^{-1}$, we have

$$g(\lambda) = h_\lambda(T)^{-1}h_\lambda(T)g(\lambda) \in Y \text{ on } D_2,$$

and hence by analytic continuation $g(\lambda) \in Y$ on D. □

2.21. *Corollary. Given* $T \in B(X)$, *let* $f:D \to C$ *be analytic on an open neighborhood* D *of* $\sigma(T)$. *If* T *has the* SVEP *then* f(T) *has that property.*

Proof. Obviously, any T has the SVEP iff the zero subspace is analytically invariant. Hence the assertion of the Corollary follows from Theorem 2.20 applied to $Y = \{0\}$. □

2.22. *Proposition. Given* $T \in B(X)$, *let* $f:G \to C$ *be analytic on an open neighborhood* G *of* $\sigma(T)$ *and nonconstant on every component of* G. *Then* T *has the* SVEP *if* f(T) *has that property.*

Proof. Suppose that T does not have the SVEP. Then there is a nonzero function $g:D \to X$ analytic on G such that

$$(\lambda - T)g(\lambda) = 0 \text{ on } D. \tag{2.9}$$

The assumption on g implies that $G \subset \sigma(T)$. For every $\lambda \in D$ there is an analytic function $h_\lambda : G \to C$ satisfying (see e.g. (1.10 and (1.11) where g_λ plays the role of h_λ)

$$f(\lambda) - f(\mu) = (\lambda - \mu)h_\lambda(\mu), \quad \lambda \in D, \ \mu \in G. \tag{2.10}$$

Applying the functional calculus to (2.10), we obtain

(2.11) $$f(\lambda)-f(T) = (\lambda-T)h_\lambda(T), \ \lambda \in D.$$

In view of (2.11), equation (2.9) becomes

$$[f(\lambda)-f(T)]g(\lambda) = 0 \text{ on } D.$$

Since f is nonconstant on $D \subset G$, there exists $\lambda_0 \in D$ such that $f'(\lambda_0) \neq 0$. Then there is a disk D' with center at λ_0 such that f^{-1} exists on f(D'). The composite function gof^{-1} is analytic on f(D') and verifies equation

$$[\mu-f(T)](gof^{-1})(\mu) = 0 \text{ on } f(D').$$

By the SVEP of f(T) it follows that

$$(gof^{-1})(\mu) = 0 \text{ on } f(D')$$

and this implies that

$$g(\lambda) = 0 \text{ on } D'.$$

By analytic continuation, we have

$$g(\lambda) = 0 \text{ on } D$$

but this contradicts the hypothesis on g. The contradiction implies that T has the SVEP. □

2.23. *Theorem. Given* $T \in B(X)$, *let* $f:D \to C$ *be a function analytic on an open neighborhood* D *of* $\sigma(T)$ *and nonconstant on every component of* D. *If* $Y \in AI[f(T)]$ *and* Y *is invariant under* T *then* $Y \in AI(T)$.

Proof. By Theorems 2.3 and 2.11, $f(T)^Y = f(T^Y)$ has the SVEP. It follows from Proposition 2.22 that T^Y has the SVEP and then Theorem 2.11 implies that $Y \in AI(T)$. □

2.24. *Definition. Given* $T \in B(X)$, $Y \in Inv(T)$ *is said to be T-absorbent if for any* $y \in Y$ *and all* $\lambda \in \sigma(T|Y)$, *the equation*

(2.12) $$(\lambda-T)x = y$$

has all solutions x *in* Y.

2.25. *Proposition. Given* $T \in B(X)$, *every T-absorbent space* Y *is a* ν*-space for* T.

Proof. If $\sigma(T|Y) \not\subset \sigma(T)$ then for some $\lambda \in \rho(T) \cap \sigma(T|Y)$,

$$R(\lambda;T)Y \not\subset Y$$

and consequently not all solutions of equation (2.12) belong to Y. □

The implication between the T-absorbent and the analytically invariant subspace is given by the following

2.26. *Theorem. If* $T \in B(X)$ *has the* SVEP *then every* T-*absorbent subspace is analytically invariant under* T.

Proof. Let Y be T-absorbent and let $f:D \to X$ be analytic and satisfy condition

$$(\lambda-T)f(\lambda) \in Y \text{ on an open } D \subset \mathbb{C}.$$

We can assume that D is connected. If $D \subset \sigma(T|Y)$ then by definition $f(\lambda) \in Y$ on D. Therefore, assume that $D \cap \rho(T|Y) \neq \emptyset$. Denote

$$g(\lambda) = (\lambda-T)f(\lambda), \ \lambda \in D \cap \rho(T|Y).$$

Since $g(\lambda) \in Y$, we can write

$$g(\lambda) = (\lambda-T)R(\lambda;T|Y)g(\lambda)$$

and then we have

$$(\lambda-T)[f(\lambda)-R(\lambda;T|Y)g(\lambda)] = 0 \text{ on } D \cap \rho(T|Y).$$

By the SVEP of T,

$$f(\lambda) = R(\lambda;T|Y)g(\lambda) \text{ on } D \cap \rho(T|Y),$$

and hence

$$f(\lambda) \in Y \text{ on } D \cap \rho(T|Y).$$

Thus it follows by analytic continuation that $f(\lambda) \in Y$ on all of D. □

The converse of this property does not hold. A counterexample for the converse is given in Appendix A.1.

There is no direct implication between analytically invariant and hyperinvariant subspaces as it can be seen from the following examples.

2.27. *Example. An analytically invariant subspace which is not hyperinvariant.*

Let $T \in B(X)$ with the SVEP have an eigenspace Z of dimension greater than 1. Each nonzero $x \in Z$ spans a one-dimensional invariant subspace $Y \subset Z$. Clearly Y is not hyperinvariant. Now let $f:D \to X$ be analytic and satisfy condition

$$(\lambda-T)f(\lambda) \in Y \text{ on an open } D \subset \mathbb{C}.$$

There is a complex-valued function g analytic on D verifying equation

$$(\lambda-T)f(\lambda) = g(\lambda)x, \ \lambda \in D. \tag{2.13}$$

Let $\alpha \in D$ be the eigenvalue for Z and hence for Y. In view of (2.13), we have

$$(\lambda-T)(\alpha-T)f(\lambda) = (\alpha-T)g(\lambda)x = g(\lambda)(\alpha-T)x = 0.$$

The SVEP of T implies

$$(\alpha-T)f(\lambda) = 0 \text{ on } D$$

and then for $\lambda \neq \alpha$, we have

$$(\lambda-\alpha)f(\lambda) = (\lambda-T)f(\lambda) \in Y.$$

Thus it follows that $Y \in AI(T)$. □

2.28. *Example. A hyperinvariant subspace which is not analytically invariant.*

Let T be the Hilbert space adjoint of the unilateral shift on $\ell^2(0,\infty)$. Then every $\lambda \in C$ with $|\lambda| < 1$ is an eigenvalue of T corresponding to the eigenvector

$$x = \sum_{n=0}^{\infty} \lambda^n e_n,$$

where $\{e_n\}$ is the natural orthonormal basis of $\ell^2(0,\infty)$. Since x spans the eigenspace E_λ, E_λ is hyperinvariant under T. Now let λ be fixed such that $|\lambda| < \frac{1}{2}$. For $\mu \in C$ arbitrary with $|\mu| < \frac{1}{2}$ put $t_0(\mu) = \mu$ and inductively write

$$t_{n+1}(\mu) = \mu t_n(\mu)-\lambda^n, \; n \geq 0.$$

We have

$$t_1(\mu) = \mu^2-1,$$
$$t_2(\mu) = \mu(\mu^2-1)-\lambda = \mu^3-\mu-\lambda,$$
$$t_3(\mu) = \mu(\mu^3-\mu-\lambda)-\lambda^2 = \mu^4-\mu^2-\mu\lambda-\lambda^2,$$
$$\dots\dots\dots\dots\dots\dots\dots\dots\dots\dots$$
$$t_k(\mu) = \mu^{k+1}-\mu^{k-1}-\mu^{k-3}\lambda^2- \dots - \lambda^{k-1}.$$

Then, for each k,

$$|t_k(\mu)| < (\tfrac{1}{2})^{k+1} + k(\tfrac{1}{2})^{k-1} < (k+1)2^{-k+1}.$$

The function defined by the series

$$x(\mu) = \sum_{n=0}^{\infty} t_n(\mu)e_n$$

is analytic on $\{\mu \in C: |\mu| < \frac{1}{2}\}$. It is seen that the range of $x(\mu)$ is not contained in E_λ but

$$(\mu-T)x(\mu) = \sum_{n=0}^{\infty} \mu t_n(\mu)e_n - \sum_{n=0}^{\infty} t_{n+1}(\mu)e_n =$$

$$= \sum_{n=0}^{\infty} [\mu t_n(\mu) - t_{n+1}(\mu)]e_n = \sum_{n=0}^{\infty} \lambda^n e_n = x \in E_\lambda .$$

Thus E_λ is not analytically invariant under T.

§ 3. *Spectral maximal spaces.*

We continue to specialize the invariant subspace so that a successful theory could be built on it.

Given $T \in B(X)$ and $F \in \mathcal{F}$, define the family of invariant subspaces

$$\mathrm{Inv}(T,F) = \{Y \in \mathrm{Inv}(T) : \sigma(T|Y) \subset F\}.$$

If Inv(T,F) is directed and has a maximal element Y then Y satisfies the condition expressed by the following

3.1. *Definition. Given* $T \in B(X)$, *an invariant subspace* Y *is called spectral maximal space of* T *if for any* $Z \in \mathrm{Inv}(T)$, *the inclusion*

$$\sigma(T|Z) \subset \sigma(T|Y) \quad \textit{implies} \quad Z \subset Y.$$

We denote by SM(T) the family of spectral maximal spaces of T.

3.2. *Proposition. Given* $T \in B(X)$, *let* $\{Y_i\}_{i \in \alpha}$ *be a family of* ν-*spaces for* T. *Then*

$$Y = \bigcap_{i \in \alpha} Y_i$$

is a ν-*space for* T *and*

$$\sigma(T|\bigcap_{i \in \alpha} Y_i) \subset \bigcap_{i \in \alpha} \sigma(T|Y_i).$$

Proof. For every $i \in \alpha$, Proposition 1.15 implies that $R(\lambda;T)Y_i \subset Y_i$ and hence

$$R(\lambda;T)Y \subset Y.$$

By Proposition 1.15, Y is a ν-space for T. Then for every $i \in \alpha$

$$Y \subset Y_i \text{ implies } \sigma(T|Y) \subset \sigma(T|Y_i). \quad \square$$

3.3. *Theorem. Given* $T \in B(X)$, *for every* $F \in \mathcal{F}$ *the subspace*

$$W = \text{c.l.m.}\ \{\bigcup Y : Y \in \mathrm{Inv}(T,F)\}$$

is hyperinvariant.

Proof. Let $x \in W$. Then x is the norm-limit of finite sums $\sum_j y_j$ with $y_j \in Y_j$ and $Y_j \in \mathrm{Inv}(T,F)$. Let $S \in B(X)$ commute with T and $\lambda \in \rho(S)$. Then for every

$Y \in \mathrm{Inv}(T,F)$ the subspace $Z = R(\lambda;S)Y$ is invariant under T. Moreover, it is easy to verify

$$T|Z = R(\lambda;S)(T|Y)(\lambda-S)|Z, \text{ for } \lambda \in \rho(S).$$

Thus the following similarity transformation holds

$$T|Z = R(\lambda;S|Z)(T|Y)(\lambda-S|Z), \text{ for } \lambda \in \rho(S|Z). \tag{3.1}$$

By (3.1),

$$\sigma(T|Z) = \sigma(T|Y).$$

Therefore, $R(\lambda;S)y_j \in W$ and since $R(\lambda;S)$ is continuous and W is closed, we have

$$Sx = \frac{1}{2\pi i}\int_\Gamma \lambda\, R(\lambda;S)x d\lambda \in W,$$

where the admissible contour $\Gamma \subset \rho(S)$ surrounds $\sigma(S)$. Thus W is invariant under S. □

3.4. *Corollary. Every spectral maximal space of* $T \in B(X)$ *is hyperinvariant.*

Proof. Clearly, if $Y \in SM(T)$ then

$$Y = \text{c.l.m.}\,\{\bigcup Z : Z \in \mathrm{Inv}\,[T,\sigma(T|Y)]\}.$$

Now for $F = \sigma(T|Y)$, Theorem 3.3 concludes the proof. □

In view of Corollary 3.4, Example 2.27 shows that not every analytically invariant subspace is spectral maximal for a given $T \in B(X)$.

3.5. *Corollary. Every spectral maximal space of* $T \in B(X)$ *is a* ν*-space for* T.

3.6. *Proposition. Given* $T \in B(X)$, *an arbitrary intersection of spectral maximal spaces of* T *is again a spectral maximal space of* T.

Proof. Let $\{Y_i\}_{i \in \alpha} \subset SM(T)$ and denote

$$Y = \bigcap_{i \in \alpha} Y_i.$$

By Corollary 3.5 and Proposition 3.2, for every $Z \in \mathrm{Inv}(T)$ with $\sigma(T|Z) \subset \sigma(T|Y)$, we have

$$\sigma(T|Z) \subset \sigma(T|\bigcap_{i \in \alpha} Y_i) \subset \bigcap_{i \in \alpha} \sigma(T|Y_i).$$

Then

$$\sigma(T|Z) \subset \sigma(T|Y_i), \text{ for all } i \in \alpha.$$

For every $i \in \alpha$, Y_i being spectral maximal, we have $Z \subset Y_i$ and hence

$$Z \subset \bigcap_{i \in \alpha} Y_i = Y. \qquad \square$$

3.7. *Theorem. Every spectral maximal space of* $T \in B(X)$, *is* T-*absorbent.*

Proof. Let $Y \in SM(T)$. Fix $y \in Y$ and $\lambda \in \sigma(T|Y)$. Assume to the contrary that there is a solution $x \notin Y$ to the equation

$$(\lambda - T)x = y.$$

The linear manifold

$$Z = \{z \in X : z = y + \alpha x,\ y \in Y,\ \alpha \in C\}$$

is closed because Y is closed. A painless verification shows that $Z \in Inv(T)$.

To obtain a contradiction, we shall ascertain that for every $\mu \in \rho(T|Y)$, $\mu - T|Z$ is bijective. Let $\mu \in \rho(T|Y)$. For $z \in Z$, the equation

$$(\mu - T)z = 0$$

implies

$$0 = (\mu - T)y + \alpha(\mu - T)x = [(\mu - T)y + \alpha(\lambda - T)x] + \alpha(\mu - \lambda)x$$

and it follows that

$$\alpha(\mu - \lambda)x \in Y.$$

Since $\mu \neq \lambda$ and $x \notin Y$, we have $\alpha = 0$ and hence

$$(\mu - T)y = 0.$$

Now $\mu \in \rho(T|Y)$ implies that $y = 0$ and hence $z = 0$.

For the proof of the surjectivity of $\mu - T|Z$ we have to find a vector

$$z' = y' + \alpha' x \in Z, \text{ with } y' \in Y \text{ and } \alpha' \in C$$

such that

$$(\mu - T)(y' + \alpha' x) = y + \alpha x. \tag{3.2}$$

Rewriting (3.2) as

$$(\mu - T)y' + \alpha'(\lambda - T)x + \alpha'(\mu - \lambda)x = y + \alpha x$$

we find that

$$(\mu - T)y' + \alpha'(\lambda - T)x = y \tag{3.3}$$

and

$$\alpha'(\mu - \lambda) = \alpha. \tag{3.4}$$

From (3.4) we obtain

$$\alpha' = \frac{\alpha}{\mu-\lambda}.$$

Since $\mu \in \rho(T|Y)$ and $y-\alpha'(\lambda-T)x \in Y$, we can solve (3.3). Thus $\mu-T|Z$ is bijective and hence

$$\sigma(T|Z) \subset \sigma(T|Y).$$

Then $Z \subset Y$ but this contradicts the assumption $x \notin Y$. □

Not every T-absorbent subspace is spectral maximal for a given $T \in B(X)$ as it can be seen from the following

3.8. Example. A T-absorbent subspace which is not spectral maximal. Let $X = C[0,1]$ denote the Banach space of continuous complex-valued functions on $[0,1]$ endowed with the norm

$$\| x \| = \sup_{[0,1]} |x(t)|, \; x \in C[0,1].$$

Define $T \in B(X)$, by

$$Tx(t) = \int_0^t x(s)ds, \; x \in X, \; t \in [0,1].$$

If $R(\lambda;T)$ exists, let denote

$$G(t) = R(\lambda;T)x(t).$$

Then

$$(\lambda-T)G(t) = x(t),$$

or

$$\lambda G(t) - \int_0^t G(s)ds = x(t). \tag{3.5}$$

For $\lambda \neq 0$, equation (3.5) has the solution

$$G(t) = \frac{1}{\lambda^2} e^{t/\lambda} \int_0^t x(s)e^{-s/\lambda}ds + \frac{1}{\lambda} x(t). \tag{3.6}$$

In order to check (3.6), perform an integration by parts on

$$\int_0^t G(s)ds = \frac{1}{\lambda^2} \int_0^t e^{s/\lambda} \int_0^s x(r)e^{-r/\lambda}\, drds + \frac{1}{\lambda} \int_0^t x(s)ds$$

to get

$$\int_0^t G(s)ds = \frac{1}{\lambda} e^{t/\lambda} \int_0^t x(s)e^{-s/\lambda}ds.$$

Then

$$(\lambda-T)R(\lambda;T)x(t) = (\lambda-T)G(t) = \lambda G(t) - \int_0^t G(s)ds = x(t).$$

Thus we have

$$\sigma(T) = \{0\}.$$

Now let

$$Y = \{x \in X: x|[0, \tfrac{1}{2}] = 0\}.$$

After a moment a reflection, we deduce that Y is a subspace invariant under both T and $R(\lambda;T)$ and hence $\sigma(T|Y) \subset \sigma(T)$. Then $\sigma(T) = \sigma(T|Y) = \{0\}$. But since $Y \neq X$, Y is not spectral maximal for T. Since $Tx \in Y$ implies $x \in Y$ by continuity, Y is T-absorbent.

3.9. *Theorem. If* $T \in B(X)$ *has the* SVEP *then every spectral maximal space of* T *is analytically invariant.*

Proof. Let $Y \in SM(T)$ and let $f: D \to X$ be analytic and verify condition

$$(\lambda-T)f(\lambda) \in Y \text{ on an open } D \subset \mathbb{C}.$$

We may suppose that D is connected. If $D \subset \sigma(T|Y)$ then Theorem 3.7 implies that $f(\lambda) \in Y$ on D. Therefore assume that

$$D \cap \rho(T|Y) \neq \emptyset .$$

For $\lambda \in D \cap \rho(T|Y)$, let

$$g(\lambda) = (\lambda-T)f(\lambda).$$

Since $g(\lambda) \in Y$, we can write

$$g(\lambda) = (\lambda-T)R(\lambda;T|Y)g(\lambda).$$

Then, by the SVEP of T,

$$f(\lambda) = R(\lambda;T|Y)g(\lambda) \text{ on } D \cap \rho(T|Y).$$

Thus $f(\lambda) \in Y$ on $D \cap \rho(T|Y)$ and on all of D by analytic continuation. □

The opposite implication does not hold in general as it can be seen in Appendix A.1.

3.10. *Corollary. Let* $T \in B(X)$ *have the* SVEP. *If* Y *is any spectral maximal space of* T *then*

$$\text{c.l.m.}\{\tilde{y}(\lambda): y \in Y, \lambda \in \rho(y,T)\} \subset Y.$$

Proof. By Theorem 3.9, $Y \in AI(T)$ and then the assertion of the Corollary follows from Corollary 2.10. □

3.11. *Theorem. Let* $T \in B(X)$ *have the* SVEP. *If for* $F \in \mathcal{F}$, $X_T(F)$ *is close*
then $X_T(F)$ *is a spectral maximal space of* T *and*

$$\sigma[T|X_T(F)] \subset F \cap \sigma(T). \tag{3.7}$$

Proof. First we prove inclusion (3.7). Let $\lambda \in F^c$. Fix an arbitrary $x \in X_T(F$
Then $\sigma(x,T) \subset F$ and by Proposition 1.5 (v), $\sigma[\tilde{x}(\lambda),T] = \sigma(x,T) \subset F$. Hence

$$(\lambda-T)\tilde{x}(\lambda) = x \tag{3.8}$$

and $\tilde{x}(\lambda) \in X_T(F)$. Equation (3.8) proves that $\lambda-T|X_T(F)$ is surjective. Since $T|X_T(F)$ has the SVEP (Proposition 1.10), Corollary 1.3 implies that $\lambda \in \rho[T|X_T(F)]$ and hence

$$\sigma[T|X_T(F)] \subset F.$$

Moreover, $X_T(F)$ being hyperinvariant (Proposition 1.8),it follows from Proposition 1.15 that

$$\sigma[T|X_T(F)] \subset \sigma(T).$$

Next, let $Y \in \mathrm{Inv}(T)$ be such that

$$\sigma(T|Y) \subset \sigma[T|X_T(F)].$$

For every $x \in Y$, with the help of (3.7) and Proposition 1.10, we obtain successively

$$\sigma(x,T) \subset \sigma(x,T|Y) \subset \sigma(T|Y) \subset \sigma[T|X_T(F)] \subset F$$

and hence $x \in X_T(F)$. Thus $Y \subset X_T(F)$ and therefore $X_T(F)$ is spectral maximal for T. □

A simple example of an operator for which $X_T(F)$ is closed for $F \in \mathcal{F}$ is the hyponormal operator T on a Hilbert space with void residual spectrum.

The local resolvent of this operator satisfies the first order growth condition

$$\| \tilde{x}(\lambda) \| \leq \frac{1}{d[\lambda,\sigma(x,T)]}, \text{ for all } \lambda \in \rho(x,T),$$

(Stampfli [2]).

3.12. *Corollary. Let* $T \in B(X)$ *have the* SVEP. *If for* $F \in \mathcal{F}$, $Y = X_T(F)$ *is closed then for every* $y \in Y$,

$$\sigma(y,T) = \sigma(y,T|Y).$$

Proof. By Theorem 3.11, $Y \in SM(T)$ and then Theorem 3.9 coupled with Proposition 2.9 concludes the proof. □

3.13. Proposition. Let $T \in B(X)$ *have the* SVEP *and let* $Y \in SM(T)$. *Then for every* $F \in \mathcal{F}$, *we have*

$$Y \cap X_T(F) = Y_{T|Y}(F).$$

Proof. Let $x \in Y \cap X_T(F)$. By Corollary 3.10, $\tilde{x}(\lambda) \in Y$ for every $\lambda \in F^c$ and therefore $\sigma(x,T|Y) \subset F$. Thus, we have

$$Y \cap X_T(F) \subset Y_{T|Y}(F).$$

Conversely, for $x \in Y_{T|Y}(F)$ Proposition 1.10 implies

$$\sigma(x,T) \subset \sigma(x,T|Y) \subset F$$

and hence $x \in X_T(F)$. Then $x \in Y \cap X_T(F)$ and

$$Y_{T|Y}(F) \subset Y \cap X_T(F). \square$$

3.14. Example. Given $T \in B(X)$, *let* τ *be a spectral set of* $\sigma(T)$ *and denote by* $E(\tau)$ *the corresponding projection. Then* $E(\tau)X$ *is a spectral maximal space of* T.

Proof. Let $Y \in \mathrm{Inv}(T)$ be such that

$$\sigma(T|Y) \subset \sigma[T|E(\tau)X] = \tau .$$

Let $y \in Y$ be arbitrary. For

$$C = \{\lambda : |\lambda| = \|T\| + 1\}$$

and for an admissible contour Γ which surrounds τ and is contained in $\tau^c \cap \rho(T)$ we have successively

$$y = \frac{1}{2\pi i}\int_C R(\lambda;T)y\,d\lambda = \frac{1}{2\pi i}\int_C R(\lambda;T|Y)y\,d\lambda =$$

$$= \frac{1}{2\pi i}\int_\Gamma R(\lambda;T|Y)y\,d\lambda = \frac{1}{2\pi i}\int_\Gamma R(\lambda;T)y\,d\lambda = E(\tau)y.$$

Hence it follows that $y \in E(\tau)X$ and consequently we have $Y \subset E(\tau)X$. $\square$

The spectral maximal spaces satisfy some transitivity properties.

3.15. Theorem. Given $T \in B(X)$ *and* $Y, Z \in \mathrm{Inv}(T)$ *with* $Y \subset Z$, *we have*

(i) *If* $Y \in SM(T)$ *then* $Y \in SM(T|Z)$;

(ii) *If* $Z \in SM(T)$ *and* $Y \in SM(T|Z)$ *then* $Y \in SM(T)$;

(iii) *If* $Y, Z \in SM(T)$ *then* $Z/Y \in SM(T^Y)$.

Proof. (i): Let $Y_1 \in \mathrm{Inv}(T|Z)$ be such that

$$\sigma[(T|Z)|Y_1] \subset \sigma[(T|Z)|Y]$$

or, equivalently,

$$\sigma(T|Y_1) \subset \sigma(T|Y).$$

Since $Y_1 \in \mathrm{Inv}(T)$ and $Y \in \mathrm{SM}(T)$, it follows that $Y_1 \subset Y$.

(ii): Let $Y_1 \in \mathrm{Inv}(T)$ with

$$\sigma(T|Y_1) \subset \sigma(T|Y).$$

Then

$$\sigma(T|Y_1) \subset \sigma(T|Z). \tag{3.9}$$

Since $Z \in \mathrm{SM}(T)$, (3.9) implies that $Y_1 \subset Z$ and hence $Y_1 \in \mathrm{Inv}(T|Z)$. We have

$$\sigma[(T|Z)|Y_1] = \sigma(T|Y_1) \subset \sigma(T|Y) = \sigma[(T|Z)|Y],$$

and since $Y \in \mathrm{SM}(T|Z)$ it follows that $Y_1 \subset Y$.

(iii): Denote the quotient space $Z/Y = \hat{Z}$ and let $\hat{Z}_1 \in \mathrm{Inv}(T^Y)$ satisfy condition

$$\sigma(T^Y|\hat{Z}_1) \subset \sigma(T^Y|\hat{Z}).$$

Putting

$$Z_1 = \{x \in X : x + Y \in \hat{Z}_1\},$$

by (i), $Y \in \mathrm{SM}(T|Z_1)$. Applying Proposition 2.2 to the operator $T|Z_1$ with the restriction $(T|Z_1)|Y$ and the coinduced $(T|Z_1)^Y$, we obtain successively:

$$\sigma(T|Z_1) = \sigma[(T|Z_1)|Y] \cup \sigma[(T|Z_1)^Y] = \sigma(T|Y) \cup \sigma(T^Y|\hat{Z}_1) \subset$$

$$\subset \sigma(T|Z) \cup \sigma(T^Y|\hat{Z}) = \sigma(T|Z) \cup \sigma[(T|Z)^Y] = \sigma(T|Z).$$

Note that (i) implies $Y \in \mathrm{SM}(T|Z)$ and then $\sigma[(T|Z)^Y] \subset \sigma(T|Z)$ follows from Proposition 1.15 (ii). Since $Z \in \mathrm{SM}(T)$ we have $Z_1 \subset Z$ and consequently $\hat{Z}_1 \subset \hat{Z}$. □

3.16. *Theorem. Let* $T \in B(X)$ *have the* SVEP *and let* F_1 *and* F_2 *be closed disjoint subsets of* $\mathbb{C}$. *If* $X_T(F_1 \cup F_2)$ *is closed then each* $X_T(F_i)$, $(i = 1,2)$ *is closed and*

$$X_T(F_1 \cup F_2) = X_T(F_1) \oplus X_T(F_2).$$

Proof. Let us denote

$$Y = X_T(F_1 \cup F_2) \text{ and } Y_i = X_T(F_i),\ i = 1,2.$$

By Theorem 3.11,

$$\sigma(T|Y) \subset F_1 \cup F_2.$$

If, for $i = 1,2, \tau_i$ is the spectral set defined by

$$\tau_i = F_i \cap \sigma(T|Y_i)$$

and $E_i = E(\tau_i)$ is the corresponding projection, then we have

$$Y = E_1Y \oplus E_2Y. \tag{3.10}$$

Let $y \in Y_i$. Then $\sigma(y,T) \subset F_i$. Also,

$$\sigma(y,T) \subset \sigma(y,T|Y_i) \subset \sigma(T|Y_i)$$

and since $y \in Y$, with the help of Corollary 3.12 we can write

$$\sigma(y,T|Y) = \sigma(y,T) \subset F_i \cap \sigma(T|Y_i) = \tau_i.$$

This implies $y \in E_iY$ and hence

$$Y \subset E_iY, \ i = 1,2. \tag{3.11}$$

Conversely, if $y \in E_iY$, then

$$\sigma(y,T) \subset \sigma[(T|Y)|E_iY] \subset \tau_i \subset F_i,$$

and hence $y \in Y_i$. Thus

$$E_iY \subset Y_i, \ i = 1,2.$$

Now, in view of (3.11) we obtain

$$E_iY = Y_i, \ i = 1,2$$

and the conclusion of the proof follows from (3.10). □

NOTES AND COMMENTS.

The single-valued extension property appeared in Dunford [2,3] and received a systematic treatment in Dunford and Schwartz [1, Part III]. It also forms a preliminary topic of Colojoară and Foiaş [3]. The proof of Proposition 1.5 (v) first appeared in Colojoară and Foiaş [1] and then in [3, 1.1.2] . Theorem 1.6 first was published in Apostol [2] and it was independently proved by Bartle and Kariotis [1] . In addition to its great significance in the local spectral theory, Theorem 1.6 is followed by the straightforward Corollary 1.7 (Bartle and Kariotis [1]) thus simplifying a more technical proof originally given by Colojoară and Foiaş [2,3]. Proposition 1.2 and Corollaries 1.3, 1.4 appeared in Finch [1].

Theorem 1.9 was proved by Sine [1]. For the more general case of non-commuting operators, Theorem 1.11 was proved by Vasilescu [1] and Corollary 1.12 by Colojoară and Foiaş [1]. Theorem 1.13 appeared in Colojoară and Foiaş [2,3], Proposition 1.14 was partially proved by Bacalu [1] and it can be retraced in J.L. Taylor [1]. Some references for Proposition 1.15 are Scroggs [1], Bartle and Kariotis [1]. Proposition 1.17 was proved by Scroggs [1].

A comparative study of ν-spaces and μ-spaces, the latter being defined by the property expressed by Proposition 2.9, was done by Bartle and Kariotis [1]. Bartle and Kariotis have the credit for Definition 2.1, Theorem 2.3, Proposition 2.5 and Proposition 2.6 [ibid.].

For the proof of Theorem 2.20 we borrowed a technique used in Colojoară and Foiaş [2,3]. For Corollary 2.21 and Proposition 2.22 we simplified the original proofs of Colojoară and Foiaş [2,3].

The concept of analytically invariant subspace (Definition 2.7) was introduced by Frunză [2] who also proved the important Theorem 2.11 [ibid.]. Proposition 2.19 is due to Bacalu [1].

The concept of T-absorbent subspace (Definition 2.24) without the restriction of being invariant under T was introduced by Vasilescu [2]. The illuminating presentation of the spectral maximal space as the maximal element of the directed family Inv(T,F) was conceived by Vasilescu [2] in an attempt to generalize the notion. The concept of spectral maximal space first appeared in Foiaş [2].

The closure M(F,T) of the linear manifold $X_T(F)$ for $T \in B(X)$ and $F \in \mathcal{F}$ appeared in Bishop [1] under the name of strong spectral manifold. $X_T(F)$ being closed for F closed represents Condition C among Dunford's sufficient conditions for an operator to be spectral (Dunford [2], Dunford and Schwartz [1, Part III]). Later, Condition C was used by Bartle [1], Bartle and Kariotis [1] and Stampfli [1,2] for localization theorems and local spectral theory of some special operators.

For Corollaries 3.4, 3.10 and Theorem 3.11 we simplified the original proofs of Foiaş [2], Colojoară and Foiaş [3]. Example 3.8 was borrowed from G. Shulberg's doctoral thesis. Theorem 3.7 is an adaptation of a lemma by Frunză [2]. Proposition 3.13 was proved by Apostol [6], Corollary 3.12, Theorem 3.15 appeared in Apostol [3] and Theorem 3.16 in Apostol [5]. Example 3.4 can be found in Colojoară and Foiaş [3, 1.3.10] .

Finally, it should be mentioned that most properties of analytically invariant subspaces were obtained in the doctoral dissertation work by Lange [2].

CHAPTER II

THE GENERAL SPECTRAL DECOMPOSITION

Everything is now prepared for the study of the spectral decomposition problem which really makes invariant subspaces important. Historically, this problem evolved from very special operators for the need of providing the self-adjoint boundary value problem in a Hilbert space with a complete orthonormal set of eigenfunctions. Dunford's extensive theory on spectral operators showed that the spectral decomposition of linear operators can go beyond the classical spectral theory of self-adjoint and normal operators.

Since then more and more general classes of linear operators which admit a certain type of spectral decomposition have been discovered. Such operators decompose the underlying space into a finite linear sum of proper invariant subspaces such that the spectrum of the given operator restricted to each invariant subspace is contained in a given subset of the complex plane. The specific property that the invariant subspaces have in common determines the type of the spectral decomposition and subsequently confines the operator to a given class from the spectral theoretic point of view.

It was found that all examples of operators known to admit a spectral decomposition have the single-valued extension property. We raised the question: Is the single-valued extension property an intrinsic element of the spectral decomposition? In order to answer this question we defined axiomatically the spectral decomposition of an operator $T \in B(X)$ in terms of unspecified invariant subspaces. Then we obtained an affirmative answer. We also found that the spectral decomposition in those general terms implies a special structure of the spectrum of the operator: the spectrum is entirely the approximate point spectrum. As a by-product, we obtained some basic elements for a functional calculus. This chapter will follow step-by-step the development of the general spectral decomposition problem by reproducing most of a paper by the authors (Erdelyi and Lange [1]).

Although the theory of general spectral decomposition can be extended to unbounded linear operators on a Banach space as well as to more general topological vector spaces, the present chapter will be confined to operators in $B(X)$.

§ 4. *Operators with spectral decomposition properties.*

We begin with a preliminary property which will give us, when necessary, an alternative way of defining the subsequent spectral decomposition.

4.1. *Proposition. Let* M *be a locally compact Hausdorff space and let* K *be a compact subset of* M. *For every open cover* $\{G_i\}_1^n$ *of* K *there is an open cover* $\{H_i\}_1^n$ *of* K *such that the sets* H_i *are relatively compact and*

$$\overline{H}_i \subset G_i, \quad i = 1,2,\ldots,n.$$

Proof. Let $\lambda \in K$ be arbitrary. Then $\lambda \in G_i$ for some i, say for $i = i_1, i_2, \ldots, i_k$. There are relatively compact open neighborhoods $V(\lambda_{i_1}), V(\lambda_{i_2}), \ldots, V(\lambda_{i_k})$ of λ such that

$$\overline{V(\lambda_{i_j})} \subset G_{i_j}, \quad j = 1,2,\ldots,k.$$

Putting

$$V_\lambda = \bigcap_{j=1}^{k} V(\lambda_{i_j})$$

we have $\overline{V_\lambda} \subset G_i$, for each G_i which contains λ. The family $\{V_\lambda\}_{\lambda\in K}$ being an open cover of a compact set K, there is a finite subcover $\{V_{\lambda_j}\}_1^m$ of K. Let

$$H_i = \bigcup\{V_{\lambda_j} : \overline{V_{\lambda_j}} \subset G_i\}, \quad i = 1,2,\ldots,n.$$

Then it follows that $\overline{H}_i \subset G_i$ and

$$K \subset \bigcup_{i=1}^{n} H_i. \quad \square$$

The spectral decomposition of the underlying space X is formulated as a linear sum of an unspecified number n of invariant subspaces. The proofs of all properties given in this chapter go through under the assumption n=2, which makes the operator more general but its relation to other spectral decompositions less suitable.

4.2. *Definition.* $T \in B(X)$ *is said to have the spectral decomposition property* (abbrev. SDP and for n=2, 2-SDP) *if for every open cover* $\{G_i\}_1^n$ of $\sigma(T)$, *there is a system* $\{Y_i\}_1^n$ *of invariant subspaces under* T *with the properties* :

(4.1)
$$\begin{cases} \text{(i)} & X = \sum_{i=1}^{n} Y_i, \\ \text{(ii)} & \sigma(T|Y_i) \subset G_i, \quad i = 1,2,\ldots,n. \end{cases}$$

Condition (ii) can be substituted by

$$\text{(ii')} \quad \sigma(T|Y_i) \subset \overline{G}_i, \quad i = 1,2,\ldots,n.$$

In fact, it is obvious that (ii) implies (ii'). Conversely, by Proposition 4.1, there is an open cover $\{H_i\}_1^n$ of $\sigma(T)$ with

$$\overline{H}_i \subset G_i, \ i = 1,2,\ldots,n.$$

Definition 4.2 implies the existence of invariant subspaces $Y_i (1 \leq i \leq n)$ which satisfy (i) and

$$\sigma(T|Y_i) \subset H_i \subset \overline{H}_i \subset G_i, \ i = 1,2,\ldots,n.$$

4.3. *Proposition. If* T *has the* SDP *and* 0 *is an isolated point of the spectrum then* T *is the sum of an invertible and a quasinilpotent operator.*

Proof. Since 0 is an isolated point of $\sigma(T)$ there is a positive integer n such that

$$\{\lambda \in C: 0 < |\lambda| < \frac{1}{n}\} \subset \rho(T).$$

Consider the following open cover of C:

$$G_1 = \{\lambda \in C: |\lambda| > \frac{1}{n+1}\}, \ G_2 = \{\lambda \in C: |\lambda| < \frac{1}{n}\}.$$

By the SDP, there are invariant subspaces Y_1, Y_2 which perform the spectral decomposition

$$X = Y_1 + Y_2,$$

$$\sigma(T|Y_i) \subset G_i, \ i = 1,2.$$

That is

$$T = T|Y_1 + T|Y_2. \tag{4.2}$$

Since $0 \in \rho(T|Y_1)$, $T|Y_1$ is invertible and $T|Y_2$ is quasinilpotent having its spectrum

$$\sigma(T|Y_2) \subset \{0\}. \ \square$$

4.4. *Lemma. Let* T *have the* SDP. *If* G *is any open set such that*

$$G \cap \sigma(T) \neq \emptyset$$

then there is a nonzero $Y \in \mathrm{Inv}(T)$ *such that*

$$\sigma(T|Y) \subset G.$$

Proof. Let H be a second open set such that G,H cover $\sigma(T)$ and $\sigma(T) \not\subset H$. By (4.1) there are $Y,Z \in \mathrm{Inv}(T)$ satisfying

(4.3) $$X = Y + Z$$

$$\sigma(T|Y) \subset G, \ \sigma(T|Z) \subset H.$$

Then $Y \neq \{0\}$ because otherwise $Z = X$ and

$$\sigma(T) = \sigma(T|Z) \subset H$$

which is impossible by the choice of H. □

This lemma has a larger range of application. For instance, it holds true if we replace (4.3) by the weaker condition

$$X = \overline{Y + Z}.$$

A generalization of a theorem by Foguel [1, Theorem 5] now follows.

4.5. *Theorem. If* T *has the* SDP *then*

$$\sigma(T) = \sigma_a(T).$$

Proof. Suppose that $\sigma(T) \neq \sigma_a(T)$. Then

$$G = [\sigma_a(T)]^c$$

is open and

$$G \cap \sigma(T) \neq \emptyset .$$

By Lemma 4.4 there is a nonzero $Y \in \text{Inv}(T)$ such that

$$\sigma(T|Y) \subset G.$$

Then there exists $\lambda \in G$ such that

$$\lambda \in \partial\sigma(T|Y) \subset \sigma_a(T|Y) \subset \sigma_a(T)$$

but this is a contradiction. □

The time has come to gather the pieces needed for the proof of the SVEP for operators with the SDP. First we have to overcome the lack of the spectral inclusion property.

4.6. *Lemma. Let* T *have the* SDP. *There is a spectral decomposition* (4.1) *with the properties*

(4.4) $$\sigma(T|Y_i) \subset \sigma(T), \ i = 1,2,\ldots,n.$$

Proof. Let $\{G_i\}_1^n$ be an open cover of $\sigma(T)$. For each bounded component V_k of $\rho(T)$ which meets G_i, choose a closed disk

$$F_{ki} \subset V_k \cap G_i .$$

If necessary, by intersecting every G_i with a relatively compact open neighborhood of $\sigma(T)$, we can have an open cover of $\sigma(T)$ such that the number of the F_{ki}'s be finite. Put

$$H_i = G_i - \bigcup_k F_{ki}, \quad i = 1,2,\ldots,n.$$

Then $\{H_i\}_1^n$ covers $\sigma(T)$ and by the SDP there is a system $\{Y_i\}_1^n \subset \mathrm{Inv}(T)$ which performs the spectral decomposition

$$X = \sum_{i=1}^n Y_i,$$

$$\sigma(T|Y_i) \subset H_i \subset G_i, \quad i = 1,2,\ldots,n.$$

In the given circumstances, Proposition 1.17 implies properties (4.4).□

4.7. *Lemma. Given* $T \in B(X)$, *let* $f:D \to X$ *be a nonzero function, analytic and satisfying condition*

$$(\lambda - T)f(\lambda) = 0 \text{ on an open } D \subset \mathbb{C}.$$

If for some open nonvoid $U \subset D$, $Y \in \mathrm{Inv}(T)$ *is such that*

$$\{f(\lambda):\lambda \in U\} \subset Y, \tag{4.5}$$

then $D \subset \sigma_p(T|Y)$.

Proof. Since for f analytic, D is locally connected , we can assume that D is connected. Define

$$H = \{\lambda \in D: f(\lambda), f'(\lambda), f''(\lambda), \ldots \in Y\}.$$

H has the following properties:

(a) $H \neq \emptyset$.

Let $\lambda_0 \in U$. For $r > 0$ sufficiently small, the circle

$$\Gamma = \{\lambda \in \mathbb{C}: |\lambda - \lambda_0| = r\} \subset U,$$

and then

$$\{f(\lambda):\lambda \in \Gamma\} \subset Y.$$

By Cauchy's formula

$$f^{(n)}(\lambda_0) = \frac{n!}{2\pi i} \int_\Gamma \frac{f(\lambda)d\lambda}{(\lambda-\lambda_0)^{n+1}},$$

we have $f^{(n)}(\lambda_0) \in Y$, $n = 0,1,2,\ldots$ where $f^{(0)}(\lambda_0)$ denotes $f(\lambda_0)$.

(b) H is open.

Let $\lambda_0 \in H$. Then $f(\lambda_0), f'(\lambda_0), \ldots \in Y$. Since $f, f', f'', \ldots$ are analytic, they admit power expansions in an open neighborhood $V(\lambda_0)$ of λ_0, and hence

$$f^{(n)}(\lambda) \in Y,\ \lambda \in V(\lambda_0),\ n = 0,1,\ldots$$

Thus it follows that

$$V(\lambda_0) \subset H.$$

(c) H is closed in D:

$$H = \bigcap_{n=0}^{\infty} [f^{(n)}]^{-1}(Y).$$

(d) $H \cap D \subset \sigma_p(T|Y)$.

Note that for every $\lambda \in D$, the vectors $f^{(n)}(\lambda)$ are not all zero because otherwise $f = 0$. Let

$$m = \min \{n : f^{(n)}(\lambda) \neq 0\}.$$

If $m = 0$,

$$Tf(\lambda) = \lambda f(\lambda),$$

and if $m > 0$,

$$Tf^{(m)}(\lambda) = \lambda f^{(m)}(\lambda).$$

In either case, for $\lambda \in D \cap H$, $f^{(m)}(\lambda)$ is an eigenvector of $T|Y$ with respect to the eigenvalue λ. Thus $\lambda \in \sigma_p(T|Y)$ and property (d) holds.

By properties (a), (b), (c), $H = D$ and then property (d) concludes the proof. □

4.8. *Lemma.* *Let* Y_1, Y_2 *be subspaces of* X *such that*

$$X = Y_1 + Y_2$$

and let $f: D \to X$ *be analytic on an open* $D \subset C$. *Then for every* $\lambda \in D$ *there is a neighborhood* $V(\subset D)$ *of* λ *and analytic functions*

$$f_i : V \to Y_i,\ i = 1,2$$

such that

$$f(\mu) = f_1(\mu) + f_2(\mu),\ \mu \in V. \tag{4.6}$$

Proof. Define the continuous map $P: Y_1 \oplus Y_2 \to Y_1 + Y_2$ by $P(y_1 \oplus y_2) = y_1 + y_2$, equipped with the norm

$$\| y_1 \oplus y_2 \| = \| y_1 \| + \| y_2 \|,\ y_i \in Y_i,\ i = 1,2.$$

P being surjective, by the open mapping theorem there is a constant $k > 0$ such that for every $x \in X$ there exists

$$y = y_1 \oplus y_2 \in Y_1 \oplus Y_2$$

satisfying conditions

$$Py = x \text{ and } \| y \| \leq k \| x \| .$$

Then every $x \in X$ can be written as

$$x = y_1 + y_2, \text{ with } \| y_1 \| + \| y_2 \| \leq k \| x \| , \; y_i \in Y_i, \; i = 1,2. \tag{4.7}$$

For λ fixed in D, let

$$f(\mu) = \sum_{n=0}^{\infty} (\mu-\lambda)^n g(n), \; g(n) \in X, \tag{4.8}$$

be the Taylor series expansion of f in a neighborhood of λ contained in D. Then, for $r > 0$ sufficiently small we have

$$V = \{\nu : |\nu-\lambda| < r\} \subset D$$

and

$$\sup_n r^n \| g(n) \| < \infty. \tag{4.9}$$

In view of (4.7), we have

$$g(n) = g_1(n) + g_2(n), \text{ with } g_i(n) \in Y_i, \; i = 1,2 \tag{4.10}$$

and

$$\| g_1(n) \| + \| g_2(n) \| \leq k \| g(n) \| , \text{ for } n = 0,1,\ldots \tag{4.11}$$

By conditions (4.9) and (4.11), for $i = 1,2$, the series

$$f_i(\mu) = \sum_{n=0}^{\infty} (\mu-\lambda)^n g_i(n) \tag{4.12}$$

converges and defines an analytic function on V with values in Y_i. Then (4.8), (4.10) and (4.12) produce (4.6).□

4.9. *Theorem. Every operator with the* SDP *has the single-valued extension property.*

Proof. Given T with the SDP, let $f: D \to X$ be analytic and verify equation

$$(\lambda-T)f(\lambda) = 0 \quad \text{on an open } D \subset C \tag{4.13}$$

We may suppose that D is connected and contained in $\sigma(T)$, since for $D \cap \rho(T) \neq \emptyset$, $f(\lambda) = 0$ on some open set and hence on all of D, by analytic continuation. Let D' be an open disk in D and let H_1 and H_2 be open half-planes covering $\sigma(T)$ such that $D' - \overline{H}_1 \neq \emptyset$. In view of Lemma 4.6, there are subspaces $Y_1, Y_2 \in \text{Inv}(T)$ which perform the following spectral decomposition:

$$X = Y_1 + Y_2,$$

(4.14) $$\sigma(T|Y_i) \subset H_i \cap \sigma(T), \; i = 1,2.$$

By Lemma 4.8, there is an open disk $V \subset D' - \overline{H}_1$ and there are analytic functions

$$f_i : V \to Y_i, \; i = 1,2$$

such that

$$f(\lambda) = f_1(\lambda) + f_2(\lambda), \text{ for all } \lambda \in V.$$

Then (4.13) implies that for $\lambda \in V$,

$$g(\lambda) = (\lambda - T)f_1(\lambda) = (T-\lambda)f_2(\lambda) \in Y_1 \cap Y_2 = Y.$$

In view of Proposition 1.15, (4.14) implies that for $\lambda \in V \subset \rho(T|Y_1)$,

$$f_1(\lambda) = R(\lambda;T|Y_1)g(\lambda).$$

Next, we propose to show that $f_1(\lambda) \in Y$ on V. Fix λ_0 in V and put $x_0 = g(\lambda_0)$. For λ with $|\lambda| > \|T\|$ we have

$$R(\lambda;T|Y_1)x_0 = R(\lambda;T)x_0 \in Y.$$

Since V lies in the unbounded component of $\rho(T|Y_1)$, by analytic continuation, we have

$$R(\lambda;T|Y_1)x_0 \in Y, \text{ for all } \lambda \in V.$$

For $\lambda = \lambda_0$,

$$R(\lambda_0;T|Y_1)x_0 = R(\lambda_0;T|Y_1)g(\lambda_0) = f_1(\lambda_0) \in Y.$$

The point λ_0 being arbitrary in V, we have $f_1(\lambda) \in Y \subset Y_2$ on V. Hence

$$f(\lambda) \in Y_2 \text{ on } V$$

and on all of D by analytic continuation.

Next, we can devise another spectral decomposition of X with respect to a couple of covering open half-planes G_1, G_2 such that

$$H_2 \cap G_2 = \emptyset \text{ and } D - \overline{G}_1 \neq \emptyset.$$

There are subspaces $Z_1, Z_2 \in \text{Inv}(T)$ which perform the spectral decomposition

$$X = Z_1 + Z_2,$$

$$\sigma(T|Z_i) \subset G_i \cap \sigma(T), \; i = 1,2.$$

Then we have

(4.15) $$\sigma(T|Y_2) \cap \sigma(T|Z_2) = \emptyset .$$

By repeating the above procedure, we find that

$$f(\lambda) \in Z_2 \text{ on } D.$$

Thus both invariant subspaces Y_2 and Z_2 satisfy hypothesis (4.5) of Lemma 4.7. Hence if $f \neq 0$, then Lemma 4.7 and relation (4.15) imply that

$$D \subset \sigma_p(T|Y_2) \cap \sigma_p(T|Z_2) = \emptyset .$$

Thus $f = 0$ and the proof is concluded. □

Now we have a decomposition of the operator with the underlying space. What about the spectrum?

4.10. Theorem. If T *has the* SDP *then for any open cover* $\{G_i\}_1^n$ *of* $\sigma(T)$, *there is a spectral decomposition* (4.1) *such that*

$$\sigma(T) = \bigcup_{i=1}^{n} \sigma(T|Y_i).$$

Proof. In view of Lemma 4.6, there is a spectral decomposition (4.1) with

$$\bigcup_{i=1}^{n} \sigma(T|Y_i) \subset \sigma(T)$$

To prove the opposite inclusion, let $x \in X$ be arbitrary but fixed. We have a representation

$$x = \sum_{i=1}^{n} y_i, \text{ with } y_i \in Y_i, \ i = 1,2,\dots,n.$$

Proposition 1.5 (i) implies

$$\sigma(x,T) \subset \bigcup_{i=1}^{n} \sigma(y_i,T) \subset \bigcup_{i=1}^{n} \sigma(T|Y_i).$$

Now, with the help of Theorem 1.9, we obtain

$$\sigma(T) = \bigcup_{x \in X} \sigma(x,T) \subset \bigcup_{i=1}^{n} \sigma(T|Y_i). \square$$

Corollary 1.3 applied to the dual operator T^* implies that if T^* has the SVEP then $\lambda - T^*$ is not surjective as long as $\lambda \in \sigma(T)$. As an application, we prove that if the original T has the SDP then a restriction of $\lambda - T^*$ is at least injective.

4.11. Corollary. If T *has the* SDP *then for every* $F \in \mathcal{F}$ *and* $\lambda \in F^c$, $(\lambda - T^*)|X_T(F^c)^{\perp}$ *is injective.*

Proof. Let $\lambda \in F^c$ and let G be open such that $F \subset G$ and $\lambda \in G^c$. Then $\{F^c,G\}$ is open cover of C. By the SDP, there are subspaces Y_1, $Y_2 \in \text{Inv}(T)$ which perform the spectral decomposition

$$X = Y_1 + Y_2$$

$$\sigma(T|Y_1) \subset F^c, \ \sigma(T|Y_2) \subset G.$$

Let $y^* \in X_T(F^c)^\perp$ verify equation

$$(\lambda-T^*)y^* = 0,$$

and let $x \in X$ be arbitrary. By the spectral decomposition, there is a representation

$$x = x_1 + x_2 \text{ with } x_i \in Y_i, \ i = 1,2.$$

Since $Y_1 \subset X_T(F^c)$, we have

$$< x_1,y^* > = 0.$$

As $\lambda \in \rho(T|Y_2)$, there is a unique $y_2 \in Y_2$ such that

$$(\lambda-T)y_2 = x_2.$$

Then

$$< x_2,y^* > = < (\lambda-T)y_2,y^* > = < y_2,(\lambda-T^*)y^* > = 0.$$

Thus, we have

$$< x,y^* > = < x_1,y^* > + < x_2,y^* > = 0.$$

Since x is arbitrary in X, it follows that $y^* = 0$.

The surjectivity of $(\lambda-T^*)|X_T(F^c)^\perp$ holds under some more restrictive conditions as it will be seen later (Proposition 12.6).

§ 5. *Operator-valued functions with spectral decomposition properties.*

In this section we shall examine the stability of the SDP under the functional calculus.

5.1. *Lemma. Let* T *have the* SDP *and let* $f:D \to C$ *be analytic on an open connected neighborhood* D *of* $\sigma(T)$. *Then* $f(T)$ *has the* SDP.

Proof. For f constant the assertion of the Lemma is obvious. We shall henceforth consider non-constant functions. Let $\{G_i\}_1^n$ be an open cover of $\sigma[f(T)]$. By the spectral mapping theorem, we have

$$f[\sigma(T)] = \sigma[f(T)] \subset \bigcup_{i=1}^{n} G_i \tag{5.1}$$

and consequently

$$\sigma(T) \subset f^{-1}\left(\bigcup_{i=1}^{n} G_i\right) = \bigcup_{i=1}^{n} f^{-1}(G_i).$$

So $\{f^{-1}(G_i)\}_1^n$ is an open cover of $\sigma(T)$. By Lemma 4.4 there is a spectral decomposition

(5.2) $$X = \sum_{i=1}^{n} Y_i$$

(5.3) $$\sigma(T|Y_i) \subset f^{-1}(G_i) \cap \sigma(T), \; i = 1,2,\dots,n.$$

In view of Proposition 1.15 the spectral inclusion property expressed by (5.3) implies that every Y_i is invariant under the resolvent and hence under $f(T)$ by the functional calculus. With the help of (5.3) we obtain successively:

$$\sigma[f(T)|Y_i] = \sigma[f(T|Y_i)] = f[\sigma(T|Y_i)] \subset f[f^{-1}(G_i)] \subset G_i, \; 1 \le i \le n.$$

Now (5.1) and (5.2) complete the spectral decomposition of $f(T)$. □

The SDP is stable under finite direct sums.

5.2. *Lemma. If* $T_i \in B(X_i)$, $(i = 1,2)$ *have the* SDP *then* $T_1 \oplus T_2$ *has that property.*

Proof. Let $\{G_i\}_1^n$ be an open cover of

(5.4) $$\sigma(T_1 \oplus T_2) = \sigma(T_1) \cup \sigma(T_2).$$

Let $\{Y_i\}_1^n \subset \mathrm{Inv}(T_1)$ and $\{Z_i\}_1^n \subset \mathrm{Inv}(T_2)$ be corresponding systems of invariant subspaces pertinent to the spectral decompositions

$$X_1 = \sum_{i=1}^{n} Y_i, \quad X_2 = \sum_{i=1}^{n} Z_i;$$

$$\sigma(T_1|Y_i) \subset G_i, \quad \sigma(T_2|Z_i) \subset G_i, \; i = 1,2,\dots,n.$$

Then $Y_i \oplus Z_i$ $(1 \le i \le n)$ are invariant subspaces of $X_1 \oplus X_2$ under $T_1 \oplus T_2$ and

$$X_1 \oplus X_2 = \sum_{i=1}^{n} (Y_i \oplus Z_i).$$

Furthermore, in view of (5.4) we have

$$\sigma[(T_1 \oplus T_2)|(Y_i \oplus Z_i)] = \sigma(T_1|Y_i) \cup \sigma(T_2|Z_i) \subset G_i, \; i = 1,2,\dots,n. \;\square$$

5.3. *Lemma. Let* T *have the* SDP *and let* τ *be a spectral set. If* $E(\tau)$ *is the corresponding spectral projection then* $T|E(\tau)X$ *has the* SDP.

Proof. Let $\{G_i\}_1^n$ be an open cover of

$$\tau = \sigma[T|E(\tau)X]$$

Without loss of generality we may assume that each G_i lies at a positive distance from $\tau' = \sigma(T)-\tau$. Let H be an open neighborhood of τ' disjoint from every G_i. Then $\{G_i \cup H\}_1^n$ covers $\sigma(T)$ and the SDP of T implies the existence of $\{Z_i\}_1^n \subset \mathrm{Inv}(T)$ satisfying the spectral decomposition

$$X = \sum_{i=1}^{n} Z_i \tag{5.5}$$

$$\sigma(T|Z_i) \subset G_i \cup H, \quad i = 1,2,\ldots,n. \tag{5.6}$$

Let $E = E(\tau)$ and $Y_i = EX \cap Z_i$. Since E commutes with T it follows from (5.5) that

$$EX = \sum_{i=1}^{n} Y_i. \tag{5.7}$$

Moreover, $EX \in SM(T)$, (Example 3.14) so

$$\sigma(T|Y_i) \subset \sigma(T|EX) = \tau.$$

Now use (5.6) and apply the functional calculus to $T|Z_i$, to find invariant subspaces V_i, W_i such that

$$Z_i = V_i \oplus W_i,$$

$$\sigma(T|V_i) \subset G_i, \ \sigma(T|W_i) \subset H, \ i = 1,2,\ldots,n.$$

Then, for every i, we have

$$Y_i = (Y_i \cap V_i) \oplus (Y_i \cap W_i) \tag{5.8}$$

so that

$$\sigma(T|Y_i \cap W_i) \subset \sigma(T|Y_i) \cap \sigma(T|W_i) \subset \tau \cap H = \emptyset, \tag{5.9}$$

by the choice of H. From (5.9) it follows that

$$Y_i \cap W_i = \{0\}.$$

Thus (5.8) implies that

$$\sigma(T|Y_i) \subset \sigma(T|V_i) \subset G_i, \ i = 1,2,\ldots,n. \tag{5.10}$$

Hence (5.7) and (5.10) prove that $T|EX$ has the SDP. □

The foregoing proof of Lemma 5.3 was prepared to be suitable for the 2-SDP case.

5.4. *Theorem. Let* T *have the* SDP. *If* $f:D \to C$ *is analytic on an open neighborhood* D *of* $\sigma(T)$ *then* $f(T)$ *has the* SDP.

Proof. Let $D_1, D_2, \ldots, D_m$ be the components of D which intersect $\sigma(T)$ and put $\tau_j = D_j \cap \sigma(T)$, $(1 \le j \le m)$. The subspaces $E(\tau_j)X$ being invariant under both T and $f(T)$, we have

$$X = \bigoplus_{j=1}^{m} E(\tau_j)X, \quad T = \bigoplus_{j=1}^{m} T|E(\tau_j)X,$$

(5.11)

$$f(T) = \bigoplus_{j=1}^{m} f(T)|E(\tau_j)X.$$

Since by Lemma 5.3 every $T|E(\tau_j)X$ has the SDP, Lemma 5.1 implies that every

$$f[T|E(\tau_j)X] = f(T)|E(\tau_j)X$$

has the SDP. Finally, Lemma 5.2 applied to the direct sum (5.11) proves that $f(T)$ has the SDP. □

In order to carry over the SDP from $f(T)$ to T we need some additional conditions on f.

5.5. *Lemma. Given* $T \in B(X)$, *let* $f:D \to C$ *be analytic and injective on an open connected neighborhood* D *of* $\sigma(T)$. *Then* T *has the* SDP *if* $f(T)$ *has that property.*

Proof. Let $\{G_i\}_1^n$ be an open cover of $\sigma(T)$. Then, we have

$$\sigma[f(T)] = f[\sigma(T)] \subset f(\bigcup_{i=1}^{n} G_i) = \bigcup_{i=1}^{n} f(G_i)$$

and hence $\{f(G_i)\}_1^n$ is an open cover of $\sigma[f(T)]$.

In view of Lemma 4.6, there is a spectral decomposition in terms of a system $\{Y_i\}_1^n \subset \mathrm{Inv}[f(T)]$, as follows

$$X = \sum_{i=1}^{n} Y_i,$$

(5.12) $$\sigma[f(T)|Y_i] \subset f(G_i) \cap \sigma[f(T)], \; i = 1,2,\ldots,n.$$

Hence every Y_i is a ν-space for $f(T)$ and then Theorem 2.4 implies that the Y_i are invariant under T. Then, with the help of (5.12) we obtain successively:

$$f[\sigma(T|Y_i)] = \sigma[f(T|Y_i)] = \sigma[f(T)|Y_i] \subset f(G_i)$$

and hence we have

$$\sigma(T|Y_i) \subset f^{-1}[f(G_i)] = G_i, \; i = 1,2,\ldots,n. \; \square$$

5.6. *Theorem. Given* $T \in B(X)$, *let* $f:D \to C$ *be analytic and injective on an open neighborhood* D *of* $\sigma(T)$. *Then* T *has the* SDP *if* $f(T)$ *has that property.*

Proof. For any spectral set τ *of* $\sigma[f(T)]$ with the corresponding spectral projection $E(\tau)$, $f(T)|E(\tau)X$ inherits the SDP from $f(T)$, by Lemma 5.3. Then the assertion of the Theorem follows via Lemmas 5.5 and 5.2 in similar lines with that of Theorem 5.4. $\square$

NOTES AND COMMENTS.

Lemma 4.8 with several applications in spectral decompositions was proved by Foiaş [3] . For an interpretation of Theorem 4.5 see Appendix A.3.

A partial isometry T satisfying the hypotheses of Proposition 4.3 turns out to be a spectral operator in Dunford's sense. The sum (4.2) represents the canonical decomposition of the spectral operator T (Dunford [3] , Dunford and Schwartz [1, XV.4.6]). Such a decomposition of a partial isometry T was obtained by Erdelyi and Miller [1] under the asymptotic condition

$$\lim_{n \to \infty} \| T^*T^n - T^nT^* \|^{1/n} = 0.$$

CHAPTER III

ASYMPTOTIC SPECTRAL DECOMPOSITIONS

How much remains true of the spectral theory if we drop the fundamental condition of linear sum decomposition? A spectral theory can be built for an operator which avails itself of finite systems of invariant subspaces with the linear sum dense in the underlying space. Thus the sum representation for the vectors in the given space can be weakened by the norm limit of sums of vectors from the invariant subspaces. We shall refer to this type of spectral theory as asymptotic spectral decomposition.

The study of this spectral theory is rewarding. Many basic properties of the spectral decomposition can be extended to the asymptotic case. Thus, for instance the spectrum of an operator which satisfies an asymptotic spectral decomposition is equal to the approximate point spectrum. Indeed, the assertion of Lemma 4.4 remains valid if we replace the linear sum of the invariant subspaces by its closure and then Theorem 4.5 follows directly.

§ 6. *Analytically decomposable operators.*

6.1. *Definition.* $T \in B(X)$ *is said to be analytically decomposable if for every open cover* $\{G_i\}_1^n$ *of* $\sigma(T)$ *there is a system of analytically invariant subspaces* $\{Y_i\}_1^n$ *performing the following asymptotic spectral decomposition*

$$(6.1)\qquad \begin{cases} X = \overline{\sum_{i=1}^{n} Y_i}\,, \\ \sigma(T|Y_i) \subset G_i \ \ (\text{or } \sigma(T|Y_i) \subset \overline{G}_i),\ 1 \le i \le n. \end{cases}$$

We shall refer to (6.1) as analytic spectral decomposition.

6.2. *Theorem. Every analytically decomposable operator has the* SVEP.

Proof. Let T be analytically decomposable and let $f: D \to X$ be analytic and verify equation

$$(\lambda - T)f(\lambda) = 0 \text{ on an open } D \subset \mathbb{C}.$$

If $f \neq 0$ then $D \cap \sigma(T) \neq \emptyset$. In this case, Lemma 4.4 applied to analytically decomposable operators, implies the existence of an analytically invariant subspace Y with

$$\sigma(T|Y) \subset D.$$

On the other hand, Lemma 2.19 implies that

$$D \subset \sigma(T|Y),$$

but this is impossible because $D \neq \emptyset$ cannot be open and compact at the same time. This contradiction implies that $f = 0$ on D. □

The extension of analytic spectral decomposition from T to $f(T)$ follows easily.

6.3. *Theorem. Given* T *analytically decomposable, let* $f: D \to \mathbb{C}$ *be analytic on an open neighborhood* D *of* $\sigma(T)$. *Then* $f(T)$ *is analytically decomposable.*

Proof. Let $\{G_i\}_1^n$ be an open cover of $\sigma[f(T)] = f[\sigma(T)]$. Since $\{f^{-1}(G_i)\}_1^n$ covers $\sigma(T)$, there is a system $\{Y_i\}_1^n \subset AI(T)$ which performs the following analytic spectral decomposition

$$X = \overline{\sum_{i=1}^{n} Y_i},$$

$$\sigma(T|Y_i) \subset f^{-1}(G_i), \; i = 1,2,\ldots,n.$$

By Theorem 2.20, each Y_i is analytically invariant under $f(T)$ and moreover, we have

$$\sigma[f(T)|Y_i] = \sigma[f(T|Y_i)] = f[\sigma(T|Y_i)] \subset f[f^{-1}(G_i)] \subset G_i, \; i = 1,2,\ldots,n. \; \square$$

6.4. *Proposition. Let* T *be analytically decomposable. If* E *is a projection commuting with* T *then* $T|EX$ *is analytically decomposable.*

Proof. Denote $S = T|EX$ and let $\{G_i\}_1^n$ be an open cover of $\sigma(S)$. Putting $G_0 = \rho(S)$, $\{G_i\}_0^n$ forms an open cover of $\sigma(T)$. There is a system $\{Y_i\}_0^n \subset AI(T)$ such that

$$X = \overline{\sum_{i=0}^{n} Y_i} \quad \text{and} \quad \sigma(T|Y_i) \subset G_i, \; i = 0,1,\ldots,n.$$

By Example 2.14, $EX \in AI(T)$ and hence the subspaces

$$Z_i = Y_i \cap EX, \; i = 0,1,\ldots,n$$

are analytically invariant under S, by Proposition 2.17 (i). Since

$$\sigma(T|Z_0) \subset \sigma(T|Y_0) \cap \sigma(T|EX) \subset G_0 \cap \sigma(S) = \emptyset,$$

$Z_0 = \{0\}$. E being a projection, we have

$$EX = \overline{\sum_{i=1}^{n} Z_i}.$$

Again, by Proposition 2.17 (i), Z_i is analytically invariant under $T|Y_i$ and hence we have

$$\sigma(S|Z_i) = \sigma(T|Z_i) \subset \sigma(T|Y_i) \subset G_i, \quad i = 1,2,\ldots,n. \ \square$$

6.5. *Theorem. Let* $T_j \in B(X_j)$, $(j = 1,2)$ *and let* $T = T_1 \oplus T_2$. *Then* T *is analytically decomposable iff each* T_j *is analytically decomposable.*

Proof. The "only if" part follows from Proposition 6.4. For the "if" part, let $\{G_i\}_1^n$ be an open cover of

$$\sigma(T) = \sigma(T_1) \cup \sigma(T_2).$$

There are analytically invariant subspaces Y_{ij} under T_j such that

$$X_j = \overline{\sum_{i=1}^{n} Y_{ij}} \text{ and } \sigma(T_j|Y_{ij}) \subset G_i, \quad i = 1,2,\ldots,n;\ j = 1,2.$$

By Proposition 2.18, for every i,

$$Y_i = Y_{i1} \oplus Y_{i2}$$

is analytically invariant under T. Since

$$X = \overline{\sum_{i=1}^{n} Y_i} \text{ and } \sigma(T|Y_i) \subset G_i, \quad i = 1,2,\ldots,n$$

T is analytically decomposable. $\square$

The property expressed by the foregoing Theorem 6.5 is valid for any finite direct sum of continuous linear operators.

6.6. *Proposition. If* T *is analytically decomposable then so is* $T|\overline{TX}$.

Proof. Write $S = T|\overline{TX}$ and let $\{G_i\}_1^n$ be an open cover of $\sigma(S)$. By Example 2.16 (ii), $\overline{TX}$ (as a special case) is analytically invariant under T and hence

$$\sigma(S) \subset \sigma(T).$$

Without loss of generality we may assume that $\{G_i\}$ covers $\sigma(T)$. T being analytically decomposable, there is a system $\{Y_i\}_1^n \subset AI(T)$ such that

$$X = \overline{\sum_{i=1}^{n} Y_i} \text{ and } \sigma(T|Y_i) \subset G_i, \quad i = 1,2,\ldots,n.$$

The $Z_i = \overline{TY_i}$ form a requisite system of analytically invariant subspaces under S, for $Z_i \in AI(T)$, by Example 2.16 (ii) and then $Z_i \in AI(S)$, by Proposition 2.17 (i). Moreover, $Z_i \in AI(T|Y_i)$ and hence we have the following inclusions

$$\sigma(S|Z_i) \subset \sigma(T|Z_i) \subset \sigma(T|Y_i) \subset G_i, \quad i = 1,2,\ldots,n.$$

Finally,

$$\overline{TX} = \overline{\sum_{i=1}^{n} Z_i}$$

follows from the continuity of T. □

Analytic spectral decompositions are stable under similarity transformations.

6.7. *Theorem. Let* T *be analytically decomposable on* X. *If for a Banach space* X_1, $S \in B(X_1)$ *is similar to* T *then* S *is analytically decomposable on* X_1.

Proof. Let $P:X \to X_1$ be a bounded invertible linear operator which performs the similarity transformation between S and T, i.e. $PT = SP$.

First, we show that if $Y \in AI(T)$ then $PY \in AI(S)$. Let $Y \in AI(T)$. Then PY is invariant under S. Let $f:D \to X_1$ be analytic and satisfy

$$(\lambda-S)f(\lambda) \in PY \text{ on } D.$$

Then

$$P^{-1}(\lambda-S)f(\lambda) \in Y \text{ on } D,$$

or

$$(\lambda-T)P^{-1}f(\lambda) \in Y \text{ on } D.$$

Since $P^{-1}f(\lambda)$ is analytic on D, by the hypothesis on Y we have $P^{-1}f(\lambda) \in Y$, and hence

$$f(\lambda) \in PY \text{ on } D.$$

Thus $PY \in AI(S)$.

Now let $\{G_i\}_1^n$ be an open cover of $\sigma(T) = \sigma(S)$ and let $\{Y_i\}_1^n \subset AI(T)$ perform the analytic spectral decomposition

$$X = \overline{\sum_{i=1}^{n} Y_i}, \quad \sigma(T|Y_i) \subset G_i, \quad i = 1,2,\ldots,n.$$

Since $S|PY_i$ is similar to $T|Y_i$ under the invertible restriction $P|Y_i$, we have

$$\sigma(S|PY_i) = \sigma(T|Y_i) \subset G_i, \quad i = 1,2,\ldots,n.$$

Also

$$\sum_{i=1}^{n} PY_i = P \sum_{i=1}^{n} Y_i$$

is dense in X_1 because P is a continuous surjection. □

Analytic spectral decompositions are stable under a perturbation by a Dunford-type scalar operator (Dunford [2.3] , Dunford and Schwartz [1, Part III]). A scalar operator S has an integral representation

$$S = \int_C \lambda dE(\lambda) \tag{6.2}$$

over the complex plane C in terms of a resolution of the identity E. Each $T \in B(X)$ which commutes with S, commutes with every element of E.

6.8. Theorem. Let T *be analytically decomposable and* S *a scalar operator. If* T *commutes with* S *then* TS *and* T + S *are analytically decomposable.*

Proof. Let E denote the resolution of the identity for S and let $\delta > 0$. Then for a suitable partition $\{b_j\}$ of $\sigma(S)$ by pairwise disjoint Borel sets and for $\lambda_j \in b_j$, we have

$$\| S-\sum_j \lambda_j E(b_j) \| \leq \delta \| T \|^{-1},$$

or

$$\| TS-\sum_j \lambda_j TE(b_j) \| \leq \delta. \tag{6.3}$$

Since each $E_j = E(b_j)$ commutes with T, every component

$$T_j = TE_j = T|E_jX$$

is analytically decomposable by Proposition 6.4. Then the direct sum operator

$$P = \sum_j \lambda_j T_j = \bigoplus_j \lambda_j T_j \tag{6.4}$$

is analytically decomposable by Theorem 6.5. By the integral representation (6.2) of S and in view of (6.3), there is a sequence $\{P_n\}$ of operators of the type P as given by (6.4) which converges to TS in the uniform operator topology.

Now let $Y \in AI(T)$. Then, by Proposition 2.17 (i), $Y_j = Y \cap E_jX$ is analytically invariant under T_j and by Proposition 2.18

$$Y = \bigoplus_j Y_j$$

is analytically invariant under P. Thus, for every n, $Y \in AI(P_n)$ and by Corollary 2.13, $Y \in AI(TS)$.

If $Z \in AI(P)$ then every $Z_j = E_j Z \in AI(T_j)$ and $Z = \bigoplus_j Z_j$ is analytically invariant under $T = \bigoplus_j T_j$, by Proposition 2.18. Hence it follows from what we have shown that $Z \in AI(TS)$.

Let $\{G_i\}$ be a finite open cover of $\sigma(TS)$. For n sufficiently large $\{G_i\}$ also covers P_n (e.g. Dunford and Schwartz [1, VII.6.3]). Since every P_n is analytically decomposable, there is a system $\{Y_i\} \subset AI(P_n)$ which satisfies

$$(6.5) \qquad X = \overline{\textstyle\sum_i Y_i}, \quad \sigma(P_n|Y_i) \subset G_i, \quad \text{for every } i \text{ and all large } n.$$

Since every $Y_i \in AI(TS)$ and for n sufficiently large we can have (rf.above cited)

$$\sigma(TS|Y_i) \subset G_i, \quad \text{for every } i,$$

we conclude that TS is analytically decomposable.

Finally, the identity

$$T + S = (\lambda + S)[(T - \lambda)R(\lambda;-S) + I], \quad \text{for } \lambda \in \rho(-S)$$

reduces the sum T+S to a product between the analytically decomposable $T-\lambda$ and the scalar operators $R(\lambda;-S)$ and $\lambda+S$. □

§ 7. *Weakly decomposable operators.*

7.1. *Definition.* $T \in B(X)$ *is called weakly decomposable if for every open cover* $\{G_i\}_1^n$ *of* $\sigma(T)$ *there is a system of spectral maximal spaces* $\{Y_i\}_1^n$ *which performs the following asymptotic spectral decomposition*

$$(7.1) \qquad \begin{cases} X = \overline{\sum_{i=1}^{n} Y_i}, \\ \sigma(T|Y_i) \subset G_i, \ (\text{or } \sigma(T|Y_i) \subset \overline{G}_i), \ i = 1,2,\dots,n. \end{cases}$$

We shall refer to (7.1) as a weak spectral decomposition.

7.2. *Proposition. Let* T *be weakly decomposable. If* $G \subset C$ *is open such that*

$$G \cap \sigma(T) \neq \emptyset$$

then there exists a nonzero spectral maximal space Y *of* T *with the property*

$$\sigma(T|Y) \subset G.$$

Proof. The assertion of the Proposition follows directly from Lemma 4.4 applied to a weak spectral decomposition. □

7.3. *Lemma. Given* $T \in B(X)$, *let* $Y \in SM(T)$ *and let* $f: D \to X$ *be a nonzero function analytic and verifying equation*

(7.2) $$(\lambda - T)f(\lambda) = 0 \text{ on an open } D \subset C.$$

Then either

$$D \cap \sigma(T|Y) = \emptyset \quad \textit{or} \quad D \subset \sigma_p(T|Y).$$

Proof. Suppose that $D \cap \sigma(T|Y) \neq \emptyset$ and let $\lambda_0 \in D \cap \sigma(T|Y)$. By differentiating (7.2) n times, we obtain

(7.3) $$Tf^{(n)}(\lambda) = \lambda f^{(n)}(\lambda) + nf^{(n-1)}(\lambda), \; n = 0,1,\ldots; \; \lambda \in D$$

where $f^{(-1)}(\lambda) = 0$. The linear manifold

$$X_n = \bigvee \{f(\lambda_0), f'(\lambda_0), \ldots, f^{(n)}(\lambda_0)\}$$

being of finite dimension is a subspace of X. X_n is invariant under T since for every

$$x = \sum_{k=0}^{n} \alpha_k f^{(k)}(\lambda_0) \in X_n, \quad (\alpha_k \in C)$$

with the help of (7.3), we obtain

$$Tx = \sum_{k=0}^{n} \alpha_k Tf^{(k)}(\lambda_0) = \sum_{k=0}^{n} \alpha_k [\lambda_0 f^{(k)}(\lambda_0) + kf^{(k-1)}(\lambda_0)] \in X_n.$$

It is easily seen that we have a triangular matrix representation for $(\lambda - T)|X_n$:

$$\begin{bmatrix} \lambda-\lambda_0 & -1 & 0 & \ldots & 0 \\ 0 & \lambda-\lambda_0 & -2 & \ldots & 0 \\ \cdots & \cdots & \cdots & \cdots & \cdots \\ 0 & 0 & 0 & & \lambda-\lambda_0 \end{bmatrix}$$

with the determinant

$$\det\,[(\lambda - T)|X_n] = (\lambda - \lambda_0)^{n+1}.$$

Thus, $R(\lambda; T|X_n)$ exists for any $\lambda \neq \lambda_0$ and therefore

$$\sigma(T|X_n) \subset \{\lambda_0\}, \; n = 0,1,\ldots$$

The hypothesis on λ_0 implies

$$\sigma(T|X_n) \subset \sigma(T|Y),$$

and since $Y \in SM(T)$, we have

$$X_n \subset Y, \; n = 0,1,\ldots$$

Thus it follows that

$$f^{(n)}(\lambda_0) \in Y, \; n = 0,1,\ldots$$

and since f is analytic, there is an open $U \subset C$ such that

$$\{f(\lambda):\lambda \in U\} \subset Y.$$

Then Lemma 4.7 implies that $D \subset \sigma_p(T|Y)$. □

7.4. *Theorem. Every weakly decomposable operator has the* SVEP.

Proof. Let T be weakly decomposable and let $f:D \to X$ be analytic and verify equation

$$(\lambda - T)f(\lambda) = 0 \text{ on an open } D \subset C.$$

We may assume that $D \subset \sigma(T)$ and is connected. By Proposition 7.2, there is a nonzero spectral maximal space Y such that

$$\sigma(T|Y) \subset D.$$

If $f \neq 0$ on D, by Lemma 7.3,

$$D \subset \sigma(T|Y).$$

Since D is open and nonvoid, this is impossible. Thus $f = 0$. □

7.5. *Corollary. Every weakly decomposable operator is analytically decomposable.*

Proof. Since every weakly decomposable operator has the SVEP, by Theorem 7.4, Theorem 3.9 implies that the spectral maximal spaces of T are analytically invariant under T. □

The stability of the weak spectral decomposition under the functional calculus is subject to a restrictive condition on f. For T weakly decomposable, Theorem 6.3, Corollary 7.5 imply that f(T) is only analytically decomposable.

7.6. *Lemma. Given* $T \in B(X)$, *let* $f:D \to C$ *be analytic and injective on an open neighborhood* D *of* $\sigma(T)$. *A subspace* Y *of* X *is spectral maximal for* T *iff it is spectral maximal for* f(T).

Proof. First, we prove the "if" part of the assertion. Let $Y \in SM[f(T)]$. Then Y is a ν-space for both f(T) and T, by Corollary 3.5 and Theorem 2.4, i.e.

$$\sigma(T|Y) \subset \sigma(T). \tag{7.4}$$

If $Z \in \mathrm{Inv}(T)$ satisfies condition

$$\sigma(T|Z) \subset \sigma(T|Y),$$

then by (7.4) and Theorem 2.4, Z is a ν-space for both T and f(T). Hence we have

$$\sigma[f(T)|Z] = \sigma[f(T|Z)] = f[\sigma(T|Z)] \subset f[\sigma(T|Y)] =$$
$$= \sigma[f(T|Y)] = \sigma[f(T)|Y]$$

and since $Y \in SM[f(T)]$, we find that $Z \subset Y$. Thus, $Y \in SM(T)$.

Conversely, let $Y \in SM(T)$ and let $Z \in Inv[f(T)]$ satisfy

$$\sigma[f(T)|Z] \subset \sigma[f(T)|Y].$$

Now Y is a ν-space for both f(T) and T, and so is Z. Then, we have

$$f[\sigma(T|Z)] = \sigma[f(T|Z)] = \sigma[f(T)|Z] \subset \sigma[f(T)|Y] = \sigma[f(T|Y)] = f[\sigma(T|Y)],$$

and hence

$$\sigma(T|Z) \subset \sigma(T|Y).$$

Since $Y \in SM(T)$, it follows that $Z \subset Y$ and hence $Y \in SM[f(T)]$. □

7.7. *Theorem. Given* $T \in B(X)$, *let* $f:D \to C$ *be analytic and injective on an open neighborhood* D *of* $\sigma(T)$. *Then* T *is weakly decomposable iff* f(T) *is weakly decomposable.*

Proof. Let f(T) be weakly decomposable and let $\{G_i\}_1^n$ be an open cover of $\sigma(T)$. Since $\sigma(T) \subset D$, the sets $H_i = G_i \cap D$ $(1 \le i \le n)$ also form an open cover of $\sigma(T)$. Then $\{f(H_i)\}_1^n$ is an open cover of $\sigma[f(T)]$ and we can find spectral maximal spaces Y_i of f(T) such that

$$X = \overline{\sum_{i=1}^{n} Y_i},$$

$$\sigma[f(T)|Y_i] \subset f(H_i),\ i = 1,2,\dots,n. \tag{7.5}$$

By Lemma 7.6, $Y_i \in SM(T)$ and from (7.5) we obtain

$$\sigma(T|Y_i) \subset H_i \subset G_i,\ i = 1,2,\dots,n.$$

Thus, T is weakly decomposable. The "only if" part of the proof is similar. □

Weak spectral decompositions are highly perishable under perturbations. Nevertheless, perturbations of weakly decomposable operators by spectral operators result in analytically decomposable operators.

7.8. *Theorem. If* T *is weakly decomposable and* Q *is quasinilpotent commuting with* T *then* T + Q *is analytically decomposable.*

Proof. Let $Y \in SM(T)$. Then Y is invariant under $T + Q$. T being weakly decomposable, it has the SVEP, hence $Y \in AI(T)$ by Theorem 3.9 and T^Y has the SVEP by Theorem 2.11. Q^Y being quasinilpotent commuting with T^Y, $T^Y + Q^Y = (T + Q)^Y$ has the SVEP by Corollary 1.12 and then Theorem 2.11 implies that Y is analytically invariant under $T + Q$. Since

$$\sigma[(T+Q)|Y] = \sigma(T|Y),$$

we see that every weak spectral decomposition for T is an analytic spectral decomposition for $T + Q$. □

7.9. *Theorem. If* T *is weakly decomposable and* A *is a spectral operator which commutes with* T *then* $T + A$ *is analytically decomposable.*

Proof. Let $A = S + Q$ be the canonical decomposition of A, where S is the scalar and Q is the quasinilpotent part of A. Then, T commutes with both S and Q, and

$$T + A = (T + Q) + S.$$

Since $T + Q$ is analytically decomposable by Theorem 7.8, so is $T + A$, by Theorem 6.8. □

7.10. *Theorem. Let* T *be weakly decomposable and let* A *be a spectral operator commuting with* T. *Then* TA *is analytically decomposable.*

Proof. Let

$$A = S + Q$$

be the canonical decomposition of A, where the scalar part commutes with the quasinilpotent Q. We have

$$TA = TS + TQ$$

where TS is analytically decomposable by Theorem 6.8 and TQ is quasinilpotent. Also TS commutes with TQ. Let E be the resolution of the identity for A.

Any spectral maximal space Y of T has the following properties:

(a) $Y \in AI(T)$, (Theorem 3.9);

(b) $Y \in AI(TS)$, (part of the proof for Theorem 6.8);

(c) $Y \in AI(TA)$, ((b) and Corollary 2.12);

(d) For any Borel set b, $Y \cap E(b)X$ is analytically invariant under $T|E(b)X$, $TS|E(b)X$ and $TA|E(b)X$ ((a), (b), (c), Example 2.14, and Proposition 2.17 (i));

(e) If for Borel sets b_j, we put $E_j = E(b_j)$, (j, finite),

$$P = \bigoplus_j \lambda_j TE_j \text{ and } Y = \bigoplus_j Y \cap E_j X$$

then $Y \in AI(P)$, ((d) and Proposition 2.18).

Now, let $\{G_i\}$ be a finite open cover of $\sigma(TA) = \sigma(TS)$ and let

$$P = \bigoplus_j \lambda_j TE_j$$

as in (6.4) with each $\lambda_j \neq 0$. Since for every j, $\lambda_j T$ is weakly decomposable, there are spectral maximal spaces Y_{ij} satisfying the weak spectral decomposition

$$X = \overline{\sum_i Y_{ij}}, \ \sigma(T|Y_{ij}) \subset \lambda_j^{-1} G_i, \text{ for every i.}$$

Then by (d), for every i, the

$$Y_i^j = Y_{ij} \cap E_j X$$

are analytically invariant under $P|E_jX$, $TS|E_jX$ and $TA|E_jX$. By putting

$$Y_i = \sum_j Y_i^j = \bigoplus_j Y_i^j ,$$

we have

$$\sigma(P|Y_i) = \bigcup_j [\lambda_j \sigma(TE_j|Y_i^j)] \subset \bigcup_j \lambda_j (\lambda_j^{-1} G_i) = G_i, \text{ for all i.}$$

By (e), (b) and (c) the Y_i are analytically invariant under P, TS and TA. Finally, we choose the approximation P of TS such that

$$\sigma(TS|Y_i) \subset G_i, \text{ whenever } \sigma(P|Y_i) \subset G_i, \text{ for every i.} \square$$

§ 8. *Spectral capacities.*

One of the most important tools in the classical spectral theory of self-adjoint operators in Hilbert spaces is the set of orthonormal projections which is extended to the resolution of the identity in Dunford's theory of spectral operators. There is an analogue of these concepts for more general operators which possess some kind of spectral theory and this is the spectral capacity.

8.1. *Definition. Given a Banach space* X *over* C, *a spectral capacity is a mapping* $E: F \to S(X)$, ($S(X)$ denotes the family of subspaces of X) *which possesses the following properties:*

(i) $E(\emptyset) = \{0\}$, $E(C) = X$;

(ii) $E(\bigcap_n F_n) = \bigcap_n E(F_n)$, *for any sequence* $\{F_n\} \subset F$;

(iii) $X = \sum_j E(\overline{G}_j)$, *for every finite open cover* $\{G_j\}$ *of* C.

We call E a *weak spectral capacity* if condition (iii) is weakened by

(iii') $X = \overline{\sum_j E(\overline{G}_j)}$, for every finite open cover $\{G_j\}$ of C.

E is said to be a 2-spectral capacity if the original condition (iii) is replaced by

(iii") $X = E(\overline{G}_1) + E(\overline{G}_2)$, for every couple of open sets G_1, G_2 which cover C.

We define the *support* of the (weak) spectral capacity E to be

$$\text{supp } E = \bigcap \{F \in F : E(F) = X\}.$$

8.2. *Definition.* $T \in B(X)$ *is said to possess a (weak) spectral capacity* E *if*

(iv) $E(F) \in \text{Inv}(T)$, *for all* $F \in F$

(v) $\sigma[T|E(F)] \subset F$, *for each* $F \in F$.

It follows from (ii) that $E(F_1) \subset E(F_2)$ whenever $F_1 \subset F_2$, for $F_1, F_2 \in F$.

The intersection property (ii) likewise holds for arbitrary families of closed sets as it can be shown with the help of Lindelöf's covering theorem.

For brevity, we shall refer to the defining properties of the spectral capacity given in Definitions 8.1,8.2 and in some pertinent remarks as (i),(ii), (iii), (iii'), (iii"), (iv), and (v) throughout this section.

8.3. *Proposition.* *Let* T *possess a weak spectral capacity* E. *If* G *is an open set such that* $G \cap \text{supp } E \neq \emptyset$ *then* $E(\overline{G}) \neq \{0\}$.

Proof. There exists a second open set H such that G and H cover C but

$$\overline{H} - \text{supp } E \neq \emptyset.$$

Then, by (iii')

$$X = \overline{E(\overline{G}) + E(\overline{H})}.$$

If $E(\overline{G}) = \{0\}$, then $E(\overline{H}) = X$ and consequently

$$X = E(\text{supp } E \cap \overline{H}).$$

But the last equality contradicts the definition of supp E. Thus $E(\overline{G}) \neq \{0\}$.

8.4. *Corollary.* *Let* T *possess a weak spectral capacity* E. *Then*

$$X = E[\sigma(T)].$$

Proof. Let $K = \text{supp } E$. We propose to show that $K \subset \sigma(T)$. Suppose that there is a $\lambda \in K - \sigma(T)$. Since $\sigma(T)$ is compact there is a closed disk F with center at λ and disjoint from $\sigma(T)$. By Proposition 8.3, $E(F) \neq \{0\}$ and then by (v),

$$\sigma[T|E(F)] \cap \sigma(T) \subset F \cap \sigma(T) = \emptyset$$

but this is clearly impossible. Hence $K \subset \sigma(T)$, and it follows that

$$X = E(K) \subset E[\sigma(T)] \subset X. \quad \square$$

8.5. *Corollary. Let* T *possess a weak spectral capacity* E. *Then*

$$\sigma[T|E(F)] \subset \sigma(T), \text{ for all } F \in \mathcal{F},$$

i.e. E(F) *is a* ν-*space for* T.

Proof. Corollary 8.4, (ii) and (v) imply

$$\sigma[T|E(F)] = \sigma[T|E(F) \cap E(\sigma(T))] = \sigma[T|E(F \cap \sigma(T))] \subset F \cap \sigma(T) \subset \sigma(T). \square$$

8.6. *Corollary. Let* $T \in B(X)$ *possess a (weak) spectral capacity* E *and let* $\{G_i\}_1^n$ *be an open cover of* $\sigma(T)$. *Then there is a system* $\{F_i\}_1^n \subset \mathcal{F}$ *such that*

$$X = \sum_{i=1}^{n} E(F_i), \text{ (resp. } X = \overline{\sum_{i=1}^{n} E(F_i)});$$

$$\overline{G}_i \subset F_i, \text{ and } \sigma[T|E(F_i)] \subset \overline{G}_i, \; i = 1,2,\ldots,n.$$

Proof. There is an open set H with the properties

$$C = [\bigcup_{i=1}^{n} G_i] \cup H = \bigcup_{i=1}^{n} (G_i \cup H) \text{ and } \overline{H} \cap \sigma(T) = \emptyset.$$

If we put $F_i = \overline{G}_i \cup \overline{H}$, (iii), (resp. (iii')) implies

$$X = \sum_{i=1}^{n} E(F_i), \text{ (resp. } X = \overline{\sum_{i=1}^{n} E(F_i)}).$$

Furthermore, (v) with the help of Corollary 8.5 implies

$$\sigma[T|E(\overline{G}_i \cup \overline{H})] \subset (\overline{G}_i \cup \overline{H}) \cap \sigma(T) = \overline{G}_i \cap \sigma(T) \subset \overline{G}_i, \; i = 1,2,\ldots,n. \square$$

8.7. *Theorem. If* $T \in B(X)$ *possesses a spectral capacity then* T *has the* SVEP.

Proof. In view of Theorem 4.9, it suffices to show that T has the SDP. If E is the spectral capacity possessed by T then Corollary 8.6 exhibits a spectral decomposition for T. $\square$

8.8. *Theorem. If* $T \in B(X)$ *possesses a spectral capacity* E *then for every* $F \in \mathcal{F}$, $E(F) \in AI(T)$.

Proof. Let $f: D \to X$ be analytic on the open connected domain D such that

$$(\lambda - T)f(\lambda) \in E(F), \text{ for all } \lambda \in D.$$

First assume that

$$D \cap F^c \neq \emptyset.$$

Since by (v),

$$\sigma[T|E(F)] \subset F,$$

there is an open disk $G \subset D \cap \rho[T|E(F)]$. For $\lambda \in G$ put

$$g(\lambda) = (\lambda-T)f(\lambda)$$

and note that

$$h(\lambda) = R[\lambda;T|E(F)]g(\lambda) \in E(F), \text{ for } \lambda \in G.$$

It follows that

$$g(\lambda) = (\lambda-T)h(\lambda), \ \lambda \in G$$

and by Theorem 8.7,

$$f(\lambda) = h(\lambda) \in E(F) \text{ on } G.$$

By analytic continuation

$$f(\lambda) \in E(F) \text{ on } D.$$

Next, assume that

$$D \subset F.$$

Let G be an open disk with $\overline{G} \subset D$ and put $K = G^c$. Since D and K^0 cover C, by (iii), we have

$$X = E(\overline{D}) + E(K).$$

By Lemma 4.8, there is an open disk $V \subset G$ and analytic functions

$$f_1:V \to E(\overline{D}), \ f_2:V \to E(K)$$

such that

$$f(\lambda) = f_1(\lambda) + f_2(\lambda) \text{ on } V.$$

Also for $\lambda \in V$,

$$(\lambda-T)f_2(\lambda) = (\lambda-T)[f(\lambda)-f_1(\lambda)] \in E(F) \cap E(\overline{D}) \subset E(F),$$

and hence

$$(\lambda-T)f_2(\lambda) \in E(K) \cap E(F) = E(K \cap F), \quad \lambda \in V.$$

Since $V \cap (K \cap F) = \emptyset$, i.e.

$$V \cap (K \cap F)^c \neq \emptyset,$$

by the first part of the proof

$$f_2(\lambda) \in E(K \cap F), \text{ for } \lambda \in V.$$

Hence for $\lambda \in V$,

$$f(\lambda) = f_1(\lambda) + f_2(\lambda) \in E(\bar{D}) + E(K \cap F) \subset E(F),$$

and by analytic continuation

$$f(\lambda) \in E(F), \text{ for all } \lambda \in D. \square$$

8.9. *Corollary. Let* T *possess a spectral capacity. Then for every open cover* $\{G_i\}_1^n$ *of* $\sigma(T)$, *there is a system* $\{Y_i\}_1^n \subset AI(T)$ *which performs the following spectral decomposition*

$$X = \sum_{i=1}^{n} Y_i,$$

$$\sigma(T|Y_i) \subset G_i, \; i = 1,2,\ldots,n.$$

Proof. The assertion of the Corollary follows from Corollary 8.6 and Theorem 8.8. $\square$

§ 9. *Decomposable spectrum.*

9.1. *Definition.* T *is said to have decomposable spectrum if for every open cover* $\{G_i\}_1^n$ *of* $\sigma(T)$, *there is an asymptotic spectral decomposition*

$$X = \overline{\sum_{i=1}^{n} Y_i}, \quad \sigma(T|Y_i) \subset G_i, \; i = 1,2,\ldots,n$$

with $\{Y_i\}_1^n \subset \mathrm{Inv}(T)$, *such that*

$$\sigma(T) = \bigcup_{i=1}^{n} \sigma(T|Y_i). \tag{9.1}$$

By Corollary 4.10, every operator with the SDP has decomposable spectrum. It is not yet known whether every operator with an asymptotic spectral decomposition has decomposable spectrum. The decomposable spectrum, however, may be a helpful spectral property.

9.2. *Theorem. Let* T *be weakly decomposable. Then the following statements are equivalent:*

(i) T *has decomposable spectrum;*

(ii) *If* $F \subset \sigma(T)$ *is closed and* $G \supset F$ *is open then there exists* $Y \in SM(T)$ *such that*

$$F \subset \sigma(T|Y) \subset G.$$

(iii) *Every system* $\{Y_i\}_1^n \subset SM(T)$ *satisfies* (9.1) *whenever*

$$X = \overline{\sum_{i=1}^{n} Y_i}. \tag{9.2}$$

Proof. Since the implication (iii) => (i) is obvious, we prove only (i) => (ii) and (ii) => (iii).

(i) => (ii): Let $F \subset \sigma(T)$ be closed and $G \supset F$ be open. Then G and F^c cover $\sigma(T)$ and hence there are Y, Z ε SM(T) satisfying conditions

$$\sigma(T|Y) \subset G,\ \sigma(T|Z) \cap F = \emptyset,\ \sigma(T) = \sigma(T|Y) \cup \sigma(T|Z).$$

Consequently, $F \subset \sigma(T|Y)$.

(ii) => (iii): Let $\{Y_i\}_1^n$ be an arbitrary system of spectral maximal spaces of T satisfying (9.2). If

$$F = \bigcup_{i=1}^{n} \sigma(T|Y_i) \neq \sigma(T),$$

then there is a Z ε SM(T) such that

$$F \subset \sigma(T|Z) \neq \sigma(T). \tag{9.3}$$

Then

$$\sigma(T|\overline{\sum_{i=1}^{n} Y_i}) \subset \bigcup_{i=1}^{n} \sigma(T|Y_i) \subset \sigma(T|Z)$$

and since Z is spectral maximal, it follows that

$$X = \overline{\sum_{i=1}^{n} Y_i} \subset Z$$

but this contradicts (9.3). Therefore, $F = \sigma(T)$. □

9.3. *Corollary. Let* T *be weakly decomposable with decomposable spectrum. If* Y, Z ε Inv(T) *are such that* $\sigma(T|Y)$ *and* $\sigma(T|Z)$ *are disjoint and both contained in* $\sigma(T)$, *then* Y + Z *is a direct sum.*

Proof. First, assume that Y,Z ε SM(T). Since $\sigma(T|Y)$ and $\sigma(T|Z)$ are compact, there exists decreasing sequences $\{G_n\}$ and $\{H_n\}$ of open sets with

$$\sigma(T|Y) = \bigcap_n G_n \quad \text{and} \quad \sigma(T|Z) = \bigcap_n H_n.$$

By Theorem 9.2 (ii), for each n there is a spectral maximal space W_n such that

$$\sigma(T|Y) \cup \sigma(T|Z) \subset \sigma(T|W_n) \subset G_n \cup H_n.$$

Then, for every n, $Y + Z \subset W_n$, and hence

$$Y + Z \subset W = \bigcap_n W_n.$$

Furthermore,

$$\sigma(T|W) \subset \bigcap_n \sigma(T|W_n) \subset \bigcap_n (G_n \cup H_n) = \sigma(T|Y) \cup \sigma(T|Z).$$

Thus $\sigma(T|W)$ is the disjoint union of two spectral sets. Let $E[\sigma(T|Y)]$ and $E[\sigma(T|Z)]$ be the corresponding projections in W. We have

$$\sigma(T|E[\sigma(T|Y)]W) \subset \sigma(T|Y), \quad \sigma(T|E[\sigma(T|Z)]W) \subset \sigma(T|Z),$$

and since $Y, Z \in SM(T)$,

$$E[\sigma(T|Y)]W \subset Y, \quad E[\sigma(T|Z)]W \subset Z.$$

Consequently,

$$W = Y \oplus Z.$$

Now let Y and Z be arbitrary invariant subspaces with $\sigma(T|Y)$ and $\sigma(T|Z)$ disjoint and contained in $\sigma(T)$. Let G and H be disjoint open neighborhoods of $\sigma(T|Y)$ and $\sigma(T|Z)$, respectively. By Theorem 9.2 (ii) there are spectral maximal spaces X_1 and X_2 of T with the properties

$$\sigma(T|Y) \subset \sigma(T|X_1) \subset G, \quad \sigma(T|Z) \subset \sigma(T|X_2) \subset H.$$

By the first part of the proof, $X_1 + X_2$ is a direct sum. Since $Y \subset X_1$ and $Z \subset X_2$, it follows that $Y + Z$ is a direct sum. □

9.4. *Theorem. Let* T *be a weakly decomposable operator with decomposable spectrum. Then* T *possesses a weak spectral capacity* $\mathcal{E}$ *such that for every* $F \in \mathcal{F}$, $\mathcal{E}(F) \in SM(T)$.

Proof. For closed $F \subset \sigma(T)$, define

$$\mathcal{E}(F) = \bigcap \{Y : Y \in SM(T),\ \sigma(T|Y) \supset F\} \tag{9.4}$$

and for arbitrary $F \in \mathcal{F}$, let

$$\mathcal{E}(F) = \mathcal{E}[F \cap \sigma(T)].$$

We propose to prove that $\mathcal{E}$ is a weak spectral capacity possessed by T.

Obviously,

$$\mathcal{E}(\emptyset) = \{0\}, \quad \mathcal{E}(C) = X$$

and for every $F \in \mathcal{F}$, $\mathcal{E}(F) \in SM(T)$, by Proposition 3.6. Moreover, for $Y \in SM(T)$ with $\sigma(T|Y) \supset F$,

$$\sigma[T|\mathcal{E}(F)] \subset \sigma(T|Y). \tag{9.5}$$

To prove (iii') of Definition 8.1, let $\{G_i\}$ be a finite open cover of $\mathbb{C}$. There is a system $\{Y_i\} \subset SM(T)$ performing the weak spectral decomposition

$$X = \overline{\textstyle\sum_i Y_i}, \quad \sigma(T|Y_i) \subset G_i \quad \text{for all } i.$$

Then, for every i,

$$Y_i \subset \bigcap \{Z : Z \in SM(T),\ \sigma(T|Z) \supset \overline{G}_i \cap \sigma(T)\} = E[\overline{G}_i \cap \sigma(T)] = E(\overline{G}_i)$$

and hence (iii') of Definition 8.1 follows.

Next, we show that

$$\sigma[T|E(F)] \subset F. \tag{9.6}$$

Let $\lambda \in F^c$ be arbitrary and let $G \supset F$ be open with $\lambda \in G^c$. By Theorem 9.2, there exists $Y \in SM(T)$ satisfying

$$F \subset \sigma(T|Y) \subset G.$$

In view of (9.5), $\lambda \in \rho[T|E(F)]$ and then (9.6) follows.

Now let $\{F_n\} \subset F$ and put

$$F = \bigcap_n F_n.$$

Then

$$E(F) \subset \bigcap_n E(F_n)$$

and since for all n, $\sigma[T|E(F_n)] \subset F_n$, we have

$$\sigma[T|\bigcap_n E(F_n)] \subset \bigcap_n \sigma[T|E(F_n)] \subset F.$$

Thus

$$\bigcap_n E(F_n) \subset Y,$$

for every $Y \in SM(T)$ with $\sigma(T|Y) \supset F$. Then it follows from (9.4) that

$$\bigcap_n E(F_n) \subset E(F).$$

and hence E satisfies (ii) of Definition 8.1. With this the proof is complete. □

9.5. *Corollary. Let* $T \in B(X)$ *be weakly decomposable with decomposable spectrum. There is a weak spectral capacity* E *possessed by* T *such that every* $Y \in SM(T)$ *has the representation*

$$Y = E[\sigma(T|Y)].$$

Proof. Let $Y \in SM(T)$ be arbitrary. Then for every weak spectral capacity E

$$\sigma(T|E[\sigma(T|Y)]) \subset \sigma(T|Y)$$

implies that

$$E[\sigma(T|Y)] \subset Y.$$

By Theorem 9.4, there is a weak spectral capacity E of the type (9.4), so that for $F = \sigma(T|Y)$, we have

$$Y \subset \bigcap\{Z : Z \in SM(T),\ \sigma(T|Z) \supset \sigma(T|Y)\} = E[\sigma(T|Y)]. \square$$

9.6. *Corollary. Let* T *be weakly decomposable with decomposable spectrum. Then there is a weak spectral capacity* E *possessed by* T *with*

$$\operatorname{supp} E = \sigma(T).$$

Proof. By Theorem 9.4 and Corollary 9.5, there exists a weak spectral capacity E possessed by T such that every $Y \in SM(T)$ has the representation $Y = E[\sigma(T|Y)]$. In view of Corollary 8.4,

$$\operatorname{supp} E \subset \sigma(T).$$

Suppose that

$$\operatorname{supp} E \neq \sigma(T),$$

and denote $G = (\operatorname{supp} E)^c$. Then $G \cap \sigma(T) \neq \emptyset$ and by Proposition 7.2, there exists a nonzero $Y \in SM(T)$ such that

$$\sigma(T|Y) \subset G.$$

We have

$$E[\sigma(T|Y)] \cap X = E[\sigma(T|Y)] \cap E(\operatorname{supp} E) = E[\sigma(T|Y) \cap \operatorname{supp} E] = \{0\}.$$

But this contradicts

$$\{0\} \neq Y = E[\sigma(T|Y)]. \square$$

9.7. *Theorem. Let* T *possess a weak spectral capacity* E. *Then in each of the following cases:*

(i) *Every* $Y \in SM(T)$ *has a representation* $Y = E[\sigma(T|Y)]$;

(ii) $\operatorname{supp} E = \sigma(T)$;

T *is weakly decomposable with decomposable spectrum.*

Proof. In view of Corollary 8.6, T is weakly decomposable. Let $\{G_i\}_1^n$ be an open cover of $\sigma(T)$ and assume to the contrary that

$$F = \bigcup_{i=1}^{n} \sigma[T|E(\overline{G}_i)] \neq \sigma(T).$$

Let $G = F^c$. Then

$$G \cap \sigma(T) \neq \emptyset.$$

In case (i), there exists a nonzero $Y \in SM(T)$ with $\sigma(T|Y) \subset G$, by Proposition 7.2. By denoting $K = \sigma(T|Y)$, in view of the representation of Y, we have

$$(9.7) \qquad E(K) \neq \{0\} \text{ with } \sigma[T|E(K)] \subset G \cap \sigma(T).$$

In case (ii), there is a closed $K \subset G \cap \sigma(T)$ with $E(K) \neq \{0\}$, by Proposition 8.3. Thus, (9.7) summarizes both cases (i) and (ii).

We may assume that $K \subset G_j$, for some j. Then $E(K) \subset E(\overline{G}_j)$, for some j and hence

$$(9.8) \qquad \sigma[T|E(K)] \subset \sigma[T|E(F)].$$

Combining (9.7) and (9.8), we obtain

$$\sigma[T|E(K)] \subset G \cap \sigma[T|E(F)] \subset G \cap F = \emptyset,$$

but then $E(K) = \{0\}$ is a contradiction. □

§ *10. Quasidecomposable operators.*

A weakly decomposable operator acquires some interesting properties if it has decomposable spectrum. A subclass of weakly decomposable operators endowed with this property contains the quasidecomposable operators. These operators are characterized by a basic property (Definition 10.1, below) of the (nonasymptotic) decomposable operators (Chapter IV).

There is now a possible doubt that the reader might have. By our trend to acquire more properties for the asymptotic spectral decompositions we may end up in the class of decomposable operators. More to the point the question is whether the class of decomposable operators is or is not a proper subclass of the quasidecomposable operators. The answer to this question was given by an ingenious example constructed by Albrecht [1] of a quasidecomposable operator which is not decomposable. This justifies our further interest in this last type of asymptotic spectral decomposition.

10.1. Definition. A weakly decomposable operator T *is said to be quasi-decomposable if* $X_T(F)$ *is closed whenever* $F \subset \mathbb{C}$ *is closed.*

An example of an operator T with the SVEP for which $X_T(F)$ is not closed for F closed is given in Colojoară and Foiaş [3, 1.3.9]. Another example of this type was given by Radjabalipour [1] .

10.2. Proposition. Let T *be quasidecomposable. Every* $Y \in SM(T)$ *has the representation*

$$(10.1) \qquad Y = X_T[\sigma(T|Y)].$$

Proof. By Theorem 3.11,

$$\sigma[T|X_T(\sigma(T|Y))] \subset \sigma(T|Y),$$

and the hypothesis on Y implies that

$$X_T[\sigma(T|Y)] \subset Y.$$

To ascertain the opposite inclusion, let $y \in Y$. It follows from Corollary 3.12, that

$$\sigma(y,T) = \sigma(y, T|Y) \subset \sigma(T|Y),$$

and hence $y \in X_T[\sigma(T|Y)]$. □

10.3. Theorem. Every quasidecomposable operator has decomposable spectrum.

Proof. Let T be quasidecomposable, $\{G_i\}$ a finite open cover of $\sigma(T)$ and let $\{Y_i\} \subset SM(T)$ perform the asymptotic spectral decomposition

$$X = \overline{\textstyle\sum_i Y_i}, \ \sigma(T|Y_i) \subset G_i, \text{ for all } i.$$

If

$$F = \bigcup_i \sigma(T|Y_i)$$

is proper in $\sigma(T)$, then $X_T(F)$ is a proper subspace in X and contains each Y_i. Then

$$X \subset X_T(F),$$

which is impossible. Hence T has decomposable spectrum. □

10.4. Theorem. Given $T \in B(X)$, *the following statements are equivalent:*

(i) T *is quasidecomposable;*

(ii) T *is weakly decomposable with decomposable spectrum such that*

(10.2) $$\sigma(x,T) = \bigcap \{\sigma(T|Y) : x \in Y, Y \in SM(T)\}, \text{ for every } x \in X;$$

(iii) T *has the* SVEP *and possesses a weak spectral capacity* $\mathcal{E}$ *such that*

(10.3) $$\sigma(x,T) = \bigcap \{F \in \mathcal{F} : x \in \mathcal{E}(F)\}, \text{ for every } x \in X.$$

Proof. (i) => (ii): In view of Theorem 10.3, it suffices to prove the containment

$$\sigma(x,T) \supset \bigcap\{\sigma(T|Y) : x \in Y, Y \in SM(T)\} = S_x.$$

This follows easily. $Z = X_T[\sigma(x,T)]$ being a spectral maximal space which contains x,

$$S_x \subset \sigma(T|Z) \subset \sigma(x,T).$$

(ii) => (iii): T has the SVEP (Theorem 7.4) and possesses a weak spectral capacity E such that $E(F)$ is spectral maximal for all $F \in \mathcal{F}$ (Theorem 9.4). Moreover, by Corollary 9.5, every $Y \in SM(T)$ has the representation

$$Y = E[\sigma(T|Y)].$$

So, for every $x \in X$, there is an $F \in \mathcal{F}$ which gives rise to the spectral maximal space

$$Y = E(F) \quad \text{with} \quad x \in E(F)$$

and conversely, to every $Y \in SM(T)$ there corresponds a closed $F = \sigma(T|Y)$ with $Y = E(F)$. Hence we have

$$\bigcap\{F \in \mathcal{F}: x \in E(F)\} = \bigcap\{\sigma(T|Y): x \in Y, Y \in SM(T)\}.$$

Then (10.3) follows from (10.2).

(iii) => (i): In view of Corollary 8.6, for every finite open cover $\{G_i\}$ of $\sigma(T)$, there corresponds a system $\{Y_i\} \subset \mathrm{Inv}(T)$ which performs the asymptotic spectral decomposition

$$X = \overline{\sum_i Y_i}, \quad \sigma(T|Y_i) \subset \overline{G}_i \quad \text{for all } i.$$

We propose to show that $X_T(F)$ is closed on $\mathcal{F}$. Clearly,

$$E(F) \subset X_T(F) \quad \text{for all} \quad F \in \mathcal{F}.$$

On the other hand, if $x \in X_T(F)$ then $\sigma(x,T) \subset F$ and by (10.3),

$$x \in E[\sigma(x,T)] \subset E(F).$$

Thus $X_T(F) \subset E(F)$ and consequently $X_T(F)$ is closed for every $F \in \mathcal{F}$.

Now, the subspaces

$$X_i = X_T[\sigma(T|Y_i)] \supset Y_i$$

form a system of spectral maximal spaces of T with the properties

$$X = \overline{\sum_i X_i},$$

$$\sigma(T|X_i) \subset \sigma(T|Y_i) \subset \overline{G}_i, \text{ for all } i.$$

NOTES AND COMMENTS.

The notion of analytically decomposable operator was introduced and studied by Lange [2] . The study of weakly decomposable operators was suggested by Colojoară and Foiaş [3], and Jafarian was the first to treat them (Jafarian [1]).

The concept of spectral capacity was introduced by Apostol [6] and the weak form by Lange [2]. Theorems 7.9 and 7.10 were proved by Lange [2]. The 2-spectral capacity is contained in an extension of the spectral capacity concept by Albrecht and Vasilescu [1].

The concepts of quasidecomposable operator and decomposable spectrum were introduced by Jafarian [1] who proved Corollary 9.3 (in a restricted form) and Theorem 10.3.

CHAPTER IV

DECOMPOSABLE OPERATORS

A class of operators with a well-developed spectral theory was introduced by Foiaş [2] under the name of decomposable operators. The decomposable operators have a satisfactory duality theory and functional calculus. They are closely related to the operators studied in Chapters II and III by uniting the properties and filling some gaps of the previous theories.

§ 11. *Properties and characterizations of decomposable operators.*

11.1. *Definition.* $T \in B(X)$ *is called decomposable if for every open cover* $\{G_i\}_1^n$ *of* $\sigma(T)$ *there is a system* $\{Y_i\}_1^n \subset SM(T)$ *performing the following spectral decomposition*

(11.1)
$$\begin{cases} X = \sum_{i=1}^{n} Y_i \\ \sigma(T|Y_i) \subset G_i \quad (\text{or } \sigma(T|Y_i) \subset \overline{G}_i), \ i = 1,2,\ldots,n. \end{cases}$$

We denote the class of decomposable operators on X by D(X). It is clear from the definition that $T \in B(X)$ is decomposable iff it has the SDP in terms of spectral maximal spaces. Thus, $T \in D(X)$ inherits from the general spectral decomposition the following properties:

(11.a) $\sigma(T) = \sigma_a(T)$;

(11.b) T has the SVEP;

(11.c) T has decomposable spectrum.

We remark that property (11.c) has a stronger meaning for decomposable operators, in the sense that *every* spectral decomposition (11.1) entails the decomposable spectrum property.

As a link to the quasidecomposable operators we have the following

11.2. *Theorem. Given* $T \in D(X)$, *for every* $F \in \mathcal{F}$, $X_T(F)$ *is closed.*

Proof. Given $F \in \mathcal{F}$, let $G \supset F$ be an arbitrary open set. Pick another open $H \subset \mathbb{C}$ satisfying

$$\sigma(T) \subset G \cup H \text{ and } F \cap \overline{H} = \emptyset .$$

There are Y_G, $Y_H \in SM(T)$ performing the spectral decomposition

(11.2)
$$X = Y_G + Y_H,$$

$$\sigma(T|Y_G) \subset G, \ \sigma(T|Y_H) \subset H.$$

Let $x \in X_T(F)$ be arbitrary. By (11.2), x has a representation

$$x = y_G + y_H \text{ with } y_G \in Y_G,\ y_H \in Y_H.$$

For $\lambda \in F^c \cap \rho(T|Y_H)$, we have

$$(\lambda - T)[\tilde{x}(\lambda) - R(\lambda;T|Y_H)y_H] = x - y_H = y_G \tag{11.3}$$

and since the function

$$\tilde{y}_G(\lambda) = \tilde{x}(\lambda) - R(\lambda;T|Y_H)y_H \tag{11.4}$$

is analytic on $F^c \cap \rho(T|Y_H)$, we have $\lambda \in \rho(y_G,T)$. Consequently,

$$\sigma(y_G,T) \subset F \cup \sigma(T|Y_H) \subset F \cup \overline{H}.$$

For an admissible contour Γ surrounding F and contained in $F^c \cap \overline{H}^c$, (11.3) and (11.4) imply

$$\frac{1}{2\pi i}\int_\Gamma \tilde{y}_G(\lambda)d\lambda = \frac{1}{2\pi i}\int_\Gamma \tilde{x}(\lambda)d\lambda - \frac{1}{2\pi i}\int_\Gamma R(\lambda;T|Y_H)y_H d\lambda . \tag{11.5}$$

By Corollary 3.10,

$$\frac{1}{2\pi i}\int_\Gamma \tilde{y}_G(\lambda)d\lambda \in Y_G,$$

and since $\overline{H}^c \subset \rho(T|Y_H)$,

$$\frac{1}{2\pi i}\int_\Gamma R(\lambda;T|Y_H)y_H d\lambda = 0.$$

Furthermore, we have

$$\frac{1}{2\pi i}\int_\Gamma \tilde{x}(\lambda)d\lambda = \frac{1}{2\pi i}\int_C R(\lambda;T)x d\lambda = x,$$

where $C = \{\lambda \in \mathbb{C}: |\lambda| = \|T\| + 1\}$. Thus (11.5) implies that $x \in Y_G$ and hence

$$X_T(F) \subset Y_G. \tag{11.6}$$

Since (11.6) holds for every $Y_G \in SM(T)$ associated with an open $G \supset F$, we have

$$X_T(F) \subset \bigcap_{G \supset F} Y_G = Y. \tag{11.7}$$

Now let $y \in Y$. Then for every $G \supset F$,

$$\sigma(y,T) = \sigma(y,T|Y_G) \subset \sigma(T|Y_G) \subset G,$$

and hence

$$\sigma(y,T) \subset \bigcap_{G \supset F} G = F.$$

Thus $y \in X_T(F)$ and hence

$$Y \subset X_T(F).$$

The latter inclusion together with (11.7) gives

$$X_T(F) = Y,$$

proving that $X_T(F)$ is closed. □

11.3. Corollary. Every decomposable operator is quasidecomposable.

$T \in D(X)$ inherits the following properties from the quasidecomposable operators:

(11.d) For every $F \in \mathcal{F}$, $X_T(F) \in SM(T)$ and

$$\sigma[T|X_T(F)] \subset F \cap \sigma(T),$$

by Theorem 3.11;

(11.e) Every spectral maximal space Y of T has the representation

$$Y = X_T[\sigma(T|Y)]$$

by Proposition 10.2.

The interesting case when (11.d) holds with equality is presented in Appendix A.2.

11.4. Corollary $T \in B(X)$ *is decomposable iff* T *has the* SDP *and* $X_T(F)$ *is closed for every* $F \in \mathcal{F}$.

Proof. Let $\{G_i\}_1^n$ be an open cover of $\sigma(T)$. There is a system $\{Y_i\}_1^n \subset \mathrm{Inv}(T)$ satisfying the general spectral decomposition

$$X = \sum_{i=1}^{n} Y_i, \quad \sigma(T|Y_i) \subset G_i, \quad i = 1,2,\ldots,n.$$

Since

$$Y_i \subset X_T[\sigma(T|Y_i)] = Z_i,$$

and

$$\sigma(T|Z_i) \subset \sigma(T|Y_i) \subset G_i, \quad i = 1,2,\ldots,n$$

T is decomposable. □

11.5. Theorem. $T \in B(X)$ *is decomposable iff for every open cover* $\{G_i\}_1^n$ *of* $\sigma(T)$ *there is a system* $\{Y_i\}_1^n \subset AI(T)$ *performing the spectral decomposition*

$$X = \sum_{i=1}^{n} Y_i, \quad \sigma(T|Y_i) \subset G_i, \quad i = 1,2,\ldots,n.$$

Proof. The "only if" part of the assertion follows from the fact that for T with the SVEP every spectral maximal space is analytically invariant (Theorem 3.9).

In view of Corollary 11.4, the converse property needs the proof that $X_T(F)$ is closed for every $F \in \mathcal{F}$. This however, has the same proof as that of Theorem 11.2 with "spectral maximal" replaced by "analytically invariant" and the reference to Corollary 3.10 replaced by the reference to Corollary 2.10.□

11.6. Corollary. If $T_j \in D(X_j)$ *for* $j = 1,2$ *then* $T = T_1 \oplus T_2 \in D(X_1 \oplus X_2)$.

Proof. Let $\{G_i\}_1^n$ be an open cover of

$$\sigma(T) = \sigma(T_1) \cup \sigma(T_2).$$

T_1 and T_2 being decomposable, there exist systems $\{Y_{i1}\}_1^n \subset SM(T_1)$ and $\{Y_{i2}\}_1^n \subset SM(T_2)$ performing the spectral decompositions

$$X_j = \sum_{i=1}^{n} Y_{ij}, \; j = 1,2;$$

$$\sigma(T|Y_{ij}) \subset G_i, \; i = 1,2,\ldots,n; \; j = 1,2.$$

By Lemma 5.2 and its proof, T has the SDP in terms of the invariant subspaces

$$Y_i = Y_{i1} \oplus Y_{i2}, \; i = 1,2,\ldots,n$$

under T. The Y_i's are analytically invariant under T (Proposition 2.18) and then Theorem 11.5 concludes the proof.□

11.7. Corollary. Given $T \in D(X)$, *let* τ *be a spectral set of* $\sigma(T)$. *If* $E(\tau)$ *is the corresponding spectral projection then* $T|E(\tau)X$ *is decomposable on* $E(\tau)X$.

Proof. By Lemma 5.3, $S = T|E(\tau)X$ has the SDP. In view of Corollary 11.4 it suffices to show that for every closed $F \subset \tau$, $Y_S(F)$ is closed, where $Y = E(\tau)X$.

Let $x \in Y_S(F)$. Since $E(\tau)X$ is spectral maximal for T (Example 3.14), Proposition 3.13 implies that $x \in X_T(F)$ and hence

$$Y_S(F) \subset X_T(F). \tag{11.8}$$

On the other hand, $\sigma(S) = \tau$ (e.g. Dunford and Schwartz [1, VII.3.20]) and therefore we have

$$Y = E(\tau)X = X_T[\sigma(S)] = X_T(\tau). \tag{11.9}$$

So, if $x \in X_T(F)$ then $\sigma(x,T) \subset F \subset \tau$ and by (11.9), $x \in X_T(\tau) = Y$. Now by Proposition 3.13, $x \in X_T(F) \cap Y = Y_S(F)$ and hence

$$X_T(F) \subset Y_S(F).$$

This coupled with (11.8) gives

$$Y_S(F) = X_T(F).$$

Since $T \in D(X)$, $X_T(F)$ is closed and then so is $Y_S(F)$.□

11.8. Theorem. $T \in B(X)$ *is decomposable iff the following conditions hold:*

(i) T *has the* SVEP;

(ii) *For any system* $\{F_i\}_1^n \subset F$ *with* $\sigma(T) \subset \bigcup_{i=1}^n F_i^0$, *the* $X_T(F_i)$ *are closed and*

$$X = \sum_{i=1}^n X_T(F_i).$$

Proof. If $T \in D(X)$ then (i) holds and Theorem 11.2 with property (11.e) prove (ii).

Conversely, assume that conditions (i) and (ii) are satisfied and let $\{G_i\}_1^n$ be an open cover of $\sigma(T)$. We can find a system $\{F_i\}_1^n$ of closed sets such that

$$\sigma(T) \subset \bigcup_{i=1}^n F_i^0 \quad \text{and} \quad F_i \subset G_i, \ i = 1,2,\ldots,n.$$

Then by Theorem 3.11, the subspaces

$$Y_i = X_T(F_i)$$

are spectral maximal for T. Thus we obtain the following spectral decomposition

$$X = \sum_{i=1}^n Y_i, \quad \sigma(T|Y_i) \subset F_i \subset G_i, \ i = 1,2,\ldots,n. \ \square$$

11.9. Theorem. $T \in B(X)$ *is decomposable iff it possesses a spectral capacity.*

Proof. If $T \in D(X)$ then for every $F \in F$,

$$E(F) = X_T(F) \tag{11.10}$$

is a spectral capacity possessed by T. Indeed, (i) and (ii) of Definition 8.1, as well as (iv) of Definition 8.2 are clearly satisfied by (11.10). Furthermore, (iii) of Definition 8.1 and (v) of Definition 8.2 follow from Theorem 11.8.

Conversely, if T possesses a spectral capacity E then Corollary 8.9 and Theorem 11.5 imply that T is decomposable. □

11.10. Proposition. *Given* $T \in D(X)$, *for every* $Y \in SM(T)$ *we have*

$$\sigma(T^Y) = \overline{\sigma(T) - \sigma(T|Y)}$$

Proof. In view of Proposition 1.14 (i), we only have to prove the inclusion

$$\sigma(T^Y) \subset \overline{\sigma(T) - \sigma(T|Y)}.$$

Suppose there is a $\lambda \in \sigma(T^Y) - \overline{\sigma(T) - \sigma(T|Y)}$. Then there is an open cover $\{G_1, G_2\}$ of $\sigma(T)$ such that

$$\overline{\sigma(T) - \sigma(T|Y)} \subset G_1, \quad \lambda \in G_1^c \quad \text{and} \quad G_2 \cap \overline{\sigma(T) - \sigma(T|Y)} = \emptyset.$$

There correspond $Y_1, Y_2 \in SM(T)$ which perform the spectral decomposition

$$X = Y_1 + Y_2, \qquad \sigma(T|Y_i) \subset G_i, \quad i = 1,2. \tag{11.11}$$

Since

$$\sigma(T|Y_2) \subset G_2 \cap \sigma(T) \subset \sigma(T|Y),$$

we have

$$Y_2 \subset Y. \tag{11.12}$$

Let $y \in Y$ have a representation

$$y = y_1 + y_2, \quad \text{with} \quad y_i \in Y_i, \quad i = 1,2.$$

On the quotient space X/Y, for $\hat{y}$ with $y \in \hat{y}$, we have $\hat{y} = \hat{y}_1$ because (11.12) implies that $\hat{y}_2 = 0$. Since $\lambda \in \rho(T|Y_1)$, there is an $x \in Y_1$ verifying equation

$$(\lambda - T)x = y_1.$$

The corresponding equation on X/Y

$$(\lambda - T^Y)\hat{x} = \hat{y}_1 = \hat{y}$$

shows that $\lambda - T^Y$ is surjective. In view of Theorems 3.9 and 2.11, T^Y has the SVEP and then Corollary 1.3 implies that $\lambda \in \rho(T^Y)$. But this contradicts the hypothesis on λ. □

11.11. Corollary. *Given* $T \in D(X)$, *for every open* $G \subset C$ *with*

$$G \cap \sigma(T) \neq \emptyset \quad \textit{and} \quad \sigma(T) \not\subset G,$$

there is a proper spectral maximal space Y *of* T *with the following properties:*

$$\sigma(T|Y) \subset \overline{G} \quad \textit{and} \quad \sigma(T^Y) \cap G = \emptyset. \tag{11.13}$$

Proof. We shall use the concept of set-spectrum (Appendix A.2). Put

$$F = \overline{G} \cap \sigma(T)$$

and denote by F^I the interior of F in the topology of $\sigma(T)$. We have $\overline{F^I} = F$ and by Corollary A.2.4 (Appendix A.2), F is a set-spectrum of T. Then for $Y = X_T(F)$, we have

$$\sigma(T|Y) = F \subset \overline{G}.$$

With the help of Proposition 11.10, we obtain successively:

$$\sigma(T^Y) = \overline{\sigma(T)-F} = \overline{\sigma(T)-[\overline{G} \cap \sigma(T)]} = \overline{\sigma(T)-\overline{G}} \subset \sigma(T) - G \subset G^c. \ \square$$

11.12. Application. Let T *have the* SDP. *If* $\sigma(T)$ *has empty interior then* T *is decomposable.*

Proof. In view of Corollary 11.4, it suffices to ascertain that $X_T(F)$ is closed for every closed $F \subset \sigma(T)$.

Let $Y \in \mathbf{Inv}(T)$ be such that $\sigma(T|Y) \subset \sigma(T)$ and let $y \in Y$. Furthermore, let V be a component of the local resolvent set $\rho(y,T)$. Since by hypothesis V cannot be contained in $\sigma(T)$, there is a disk $D \subset V \cap \rho(T)$. By Proposition 1.15,

$$R(\lambda;T)y \in Y \text{ on } D$$

and by analytic continuation to all of V, the range of $\tilde{y}(\lambda)$ lies in Y, i.e.

$$\{\tilde{y}(\lambda) : \lambda \in \rho(y,T)\} \subset Y. \tag{11.14}$$

To prove that $X_T(F)$ is closed for F closed, apply the proof of Theorem 11.2 with the following modifications:

a) consider Y_G and Y_H just invariant subspaces instead of spectral maximal spaces;

b) the reference to Corollary 3.10 be replaced by the reference to property (11.14) above. $\square$

Some basic results of the functional calculus on operators with the SDP have a straightforward application to the class of decomposable operators.

11.13. Theorem. Given $T \in D(X)$, *let* $f: D \to C$ *be analytic on an open neighborhood* D *of* $\sigma(T)$. *Then* $f(T) \in D(X)$.

Proof. For f constant, f(T) is obviously decomposable. We assume therefore that f is a nonconstant function

By Theorem 5.4, f(T) has the SDP. Moreover, by Corollary 1.7, for every $F \in \mathcal{F}$ we have

$$X_{f(T)}(F) = X_T[f^{-1}(F)]. \tag{11.15}$$

Since $f^{-1}(F)$ is closed and T is decomposable, it follows from (11.15) that $X_{f(T)}(F)$ is closed. The proof is now concluded by Corollary 11.4. $\square$

11.14. *Theorem. Given* $T \in B(X)$, *let* $f: D \to C$ *be analytic and injective on an open neighborhood* D *of* $\sigma(T)$. *Then* T *is decomposable if so is* $f(T)$.

Proof. By Theorem 5.6, T has the SDP and then for every $F \in \mathcal{F}$, Corollary 1.7 implies that

$$X_T(F) = X_{f(T)}[f(T)].$$

Thus $X_T(F)$ is closed and then Corollary 11.4 concludes the proof. □

The injectivity condition on f is strongly restrictive. Apostol [6] showed that the implication $f(T) \in D(X) \Rightarrow T \in D(X)$ holds under a different restrictive condition on f. That is, if f is locally nonconstant on $\sigma(T)$, (i.e. if the zeros of f' have no accumulation point in $\sigma(T)$) then $T \in D(X)$ if $f(T) \in D(X)$.

§ 12. *The duality theory of spectral decompositions.*

For the theory we shall develop we need both a strengthening and a weakening of the decomposable operator concept. We shall reach the best result of this theory when we conclude that both modifications of the basic Definition 11.1 are insubstantial. We begin with the presentation of the stronger concept.

12.1. *Definition.* $T \in B(X)$ *is called strongly decomposable if for any open cover* $\{G_i\}_1^n$ *of* $\sigma(T)$ *and for every* $Y \in SM(T)$, *there is a system* $\{Y_i\}_1^n \subset SM(T)$ *which gives rise to the following spectral decomposition*

$$(12.1)\qquad \begin{cases} Y = \sum_{i=1}^{n} Y \cap Y_i, \\ \sigma(T|Y_i) \subset G_i, \quad i = 1,2,\ldots,n. \end{cases}$$

We shall refer to (12.1) as a strong spectral decomposition.

12.2. *Theorem.* $T \in B(X)$ *is strongly decomposable iff for any* $Y \in SM(T)$, $T|Y$ *is decomposable.*

Proof. First assume that T is strongly decomposable. Let $Y \in SM(T)$ and let $\{G_i\}_1^n$ be an open cover of $\sigma(T|Y)$. Choose an open set G_0 such that $\{G_i\}_0^n$ covers $\sigma(T)$ and $G_0 \cap \sigma(T|Y) = \emptyset$. Let $\{Y_i\}_0^n \subset SM(T)$ which performs the following strong spectral decomposition:

$$Y = \sum_{i=0}^{n} Y \cap Y_i, \quad \sigma(T|Y_i) \subset G_i, \quad i = 0,1,\ldots,n.$$

The subspaces $Z_i = Y \cap Y_i$ $(0 \le i \le n)$ are spectral maximal for $T|Y$ (Proposition 3.6 and Theorem 3.15 (i)) and verify properties:

$$\sigma[(T|Y)|Z_0] \subset \sigma(T|Y_0) \cap \sigma(T|Y) \subset G_0 \cap \sigma(T|Y) = \emptyset, \tag{12.2}$$

$$\sigma[(T|Y)|Z_i] \subset \sigma(T|Y_i) \subset G_i, \quad i = 1,2,\ldots,n. \tag{12.3}$$

Inclusions (12.2) imply that $Z_0 = \{0\}$. Furthermore, the strong decomposability of T implies

$$Y = \sum_{i=0}^{n} Y \cap Y_i = \sum_{i=0}^{n} Z_i = \sum_{i=1}^{n} Z_i. \tag{12.4}$$

The decomposition (12.4) coupled with (12.3) prove that $T|Y \in D(Y)$. Actually, $T|Y$ is strongly decomposable itself. In fact, if $Z \in SM(T|Y)$ then $Z \in SM(T)$ by Theorem 3.15 (ii). T being strongly decomposable and since $Z_0 = \{0\}$, we have

$$\sum_{i=1}^{n} Z \cap Z_i = \sum_{i=1}^{n} (Z \cap Y) \cap Y_i = Z \cap Y = Z.$$

Next, assume that for any spectral maximal space Y of T, $T|Y$ is decomposable. In particular, for $Y = X$, T is decomposable and hence it has the SVEP. Let $\{G_i\}_1^n$ be an open cover of $\sigma(T)$ and let $\{F_i\}_1^n \subset F$ have the following properties

$$F_i \subset G_i \quad \text{and} \quad \bigcup_{i=1}^{n} F_i^0 \supset \sigma(T).$$

Putting $F = \sigma(T|Y)$ and $Y_i = X_T(F_i)$, we have $Y = X_T(F)$ and with the help of Proposition 3.13, we obtain

$$\sum_{i=1}^{n} Y \cap Y_i = \sum_{i=1}^{n} Y \cap X_T(F_i) = \sum_{i=1}^{n} Y_{T|Y}(F_i) = Y. \tag{12.5}$$

The $Y_{T|Y}(F_i)$ being spectral maximal spaces for $T|Y$ and hence for T, we have

$$\sigma(T|Y_i) \subset F_i \subset G_i, \quad i = 1,2,\ldots,n. \tag{12.6}$$

Relations (12.5) and (12.6) prove that T is strongly decomposable.□

12.3. *Lemma. Let* T *be strongly decomposable and let* $Y \in SM(T)$. *If* $\hat{Z} \in SM(T^Y)$ *then*

$$Z = \{x \in X: \hat{x} = x + Y \in \hat{Z}\}$$

is spectral maximal for T.

Proof. The assertion of the Lemma will follow from

$$Z = X_T[\sigma(T|Z)]. \tag{12.7}$$

Since $Z \in \mathrm{Inv}(T)$, we have

$$\sigma(T|Z) \subset \sigma(T|X_T[\sigma(T|Z)])$$

and since $X_T[\sigma(T|Z))] \in SM(T)$, it follows that

$$Z \subset X_T[\sigma(T|Z)].$$

For convenience, let us write

$$W = X_T[\sigma(T|Z)].$$

Having $Y \subset Z \subset W$, Proposition 11.10 applied to $T|W$ gives

$$\sigma[(T|W)^Y] = \overline{\sigma(T|W) - \sigma(T|Y)} \subset \overline{\sigma(T|Z) - \sigma(T|Y)}. \tag{12.8}$$

Since $Y \in SM(T|Z)$ by Theorem 3.15 (i), Proposition 2.2 applied to $T|Z$ with restriction $(T|Z)|Y = T|Y$ and coinduced $(T|Z)^Y$ gives

$$\sigma(T|Z) = \sigma(T|Y) \cup \sigma[(T|Z)^Y].$$

Then (12.8) becomes

$$\sigma[(T|W)^Y] \subset \sigma[(T|Z)^Y] ,$$

or, equivalently

$$\sigma(T^Y|\hat{W}) \subset \sigma(T^Y|\hat{Z}).$$

Since $\hat{Z} \in SM(T^Y)$, it follows that $\hat{W} \subset \hat{Z}$, and hence

$$X_T[\sigma(T|Z)] = W \subset Z. \square$$

12.4. Theorem. If T *is strongly decomposable then for every* $Y \in SM(T)$, T^Y *is strongly decomposable.*

Proof. Let $\{G_i\}_1^n$ be an open cover of $\sigma(T^Y)$ and let G_0 be open such that $\{G_i\}_0^n$ covers $\sigma(T)$ and $G_0 \cap \sigma(T^Y) = \emptyset$. Let $\{Y_i\}_0^n \subset SM(T)$ perform the following strong spectral decomposition

$$Y = \sum_{i=0}^{n} Y \cap Y_i, \quad \sigma(T|Y_i) \subset G_i, \ i = 0,1,\dots,n.$$

Put

$$\tau_i = \sigma(T|Y_i) \cup \sigma(T|Y) \text{ and } Z_i = X_T(\tau_i), \ i = 0,1,\dots,n.$$

Then the $\hat{Z}_i = Z_i/Y \in SM(T^Y)$ by Theorem 3.15 (iii). Proposition 11.10 applied to $T|Z_i$ gives

(12.9) $\sigma(T^Y|\hat{Z}_i) = \sigma[(T|Z_i)^Y] = \overline{\sigma(T|Z_i) - \sigma(T|Y)} \subset \sigma(T|Y_i) \subset G_i$, $i = 1,2,\ldots,n$.

As for $i = 0$, with the help of Proposition 1.14 (i) we obtain

$$\sigma(T|Y_0) \subset \sigma(T) \cap G_0 = [\sigma(T|Y) \cup \sigma(T^Y)] \cap G_0 \subset \sigma(T|Y),$$

and since $Y \in SM(T)$, it follows that $Y_0 \subset Y$. Consequently, we have

$$Z_0 = X_T(\tau_0) = X_T[\sigma(T|Y)] = Y$$

and hence $\hat{Z}_0 = \{0\}$. Furthermore, for every i,

$$Z_i = X_T[\sigma(T|Y_i) \cup \sigma(T|Y_0)] \supset X_T[\sigma(T|Y_i)] = Y_i.$$

Let $\hat{Z} \in SM(T^Y)$. Then by Lemma 12.3,

$$Z = \{x \in X: \hat{x} = x + Y \in \hat{Z}\} \in SM(T)$$

and since T is strongly decomposable, we have

$$Z = \sum_{i=0}^{n} Z \cap Y_i \subset \sum_{i=0}^{n} Z \cap Z_i.$$

This, on the quotient space X/Y corresponds to

$$\hat{Z} = \sum_{i=1}^{n} \hat{Z} \cap \hat{Z}_i$$

and with (12.9) proves that T^Y is strongly decomposable. □

Now we shall weaken the concept of decomposable operator.

12.5. Definition. $T \in B(X)$ *is called 2-decomposable if for every couple of open sets* G_1, G_2 *which cover* C, *there are spectral maximal spaces* Y_1, Y_2 *performing the spectral decomposition*

$$X = Y_1 + Y_2$$

(12.10)

$$\sigma(T|Y_i) \subset G_i, \quad i = 1,2.$$

In Definition 12.5 we can also use a 2-member open cover $\{G_1, G_2\}$ of $\sigma(T)$. Indeed, we can choose an open set H such that

$$C = (G_1 \cup H) \cup (G_2 \cup H) \text{ and } H \cap \sigma(T) = \emptyset.$$

Then, there are $Y_1, Y_2 \in SM(T)$ such that

$$X = Y_1 + Y_2$$

and

$$\sigma(T|Y_i) \subset (G_i \cup H) \cap \sigma(T) \subset G_i, \quad i = 1,2.$$

As a straightforward consequence of Definition 12.5, every 2-decomposable operator T has the following properties:

(12.a) T has the SVEP (by the 2-SDP);

(12.b) For every $F \in \mathcal{F}$, $X_T(F) \in SM(T)$ with

$$\sigma[T|X_T(F)] \subset F \cap \sigma(T)$$

and every $Y \in SM(T)$ has the representation

$$Y = X_T[\sigma(T|Y)]$$

(note that the proof of Theorem 11.2 is based on a spectral decomposition of type (12.10));

(12.c) T is 2-decomposable iff T possesses a 2-spectral capacity.

We can now proceed toward the dual theory of spectral decompositions.

12.6. *Proposition. Let* T *be a 2-decomposable operator. Then for every* $F \in \mathcal{F}$,

$$\sigma[T^*|X_T(F^c)^\perp] \subset F. \tag{12.11}$$

Proof. Let $\lambda \in F^c$. We show that $(\lambda-T^*)|X_T(F^c)^\perp$ is bijective. It is injective by Corollary 4.11. Let $G \supset F$ be open such that $\lambda \in G^c$. Then $\{F^c, G\}$ is an open cover of $\mathbb{C}$. There are $Y_1, Y_2 \in SM(T)$ which decompose X:

$$X = Y_1 + Y_2,\ \sigma(T|Y_1) \subset F^c,\ \sigma(T|Y_2) \subset G.$$

Note that $Y_1 \subset X_T(F^c)$ and $Y_2 \subset X_T(G)$. Let $u \in X_T(F^c)^\perp$ be arbitrary. Fix an arbitrary x in X and consider the representation

$$x = y_1 + y_2, \text{ with } y_i \in Y_i,\ i = 1,2. \tag{12.12}$$

Define the linear functional v by

$$< x,v > = < R(\lambda;T|Y_2)y_2,\ u >.$$

This definition does not depend on a particular representation of x. In fact, let

$$x = y_1' + y_2' \text{ with } y_i' \in Y_i,\ i = 1,2$$

be another representation of x. We have

$$y_2' - y_2 = y_1 - y_1' \in Y_1 \cap Y_2.$$

Since $Y_1 \cap Y_2 \in SM(T)$, it is invariant under $R(\lambda;T|Y_2)$,

$$R(\lambda;T|Y_2)(y_2' - y_2) \in Y_1 \cap Y_2 \subset X_T(F^c)$$

and then

$$< R(\lambda;T|Y_2)(y_2' - y_2),u > = 0.$$

Thus, v is well defined :

$$< R(\lambda;T|Y_2)y_2',u > = < R(\lambda;T|Y_2)y_2,u > .$$

Next, we show that v is bounded. The linear map

$$y_1 \oplus y_2 \to y_1 + y_2$$

being continuous and surjective, it follows from the open mapping theorem that there is a constant K such that

$$\| y_1\| + \| y_2 \| \leq K \| y_1 + y_2 \| = K \| x \| .$$

Then we have

$$|< x,v >| = |< R(\lambda;T|Y_2)y_2,u >| \leq \| R(\lambda;T|Y_2) \| \cdot \| y_2\| \cdot \| u \| \leq$$

$$\leq [K \| R(\lambda;T|Y_2) \| \cdot \| u \|] \| x \| .$$

To see that $v \in X_T(F^c)^{\perp}$, let $x \in X_T(F^c)$ be arbitrary. Then, from

$$x = y_1 + y_2 \text{ with } y_i \in Y_i, \ i = 1,2$$

we obtain

$$y_2 = x-y_1 \in X_T(F^c) \cap Y_1 \subset X_T(F^c),$$

and since $X_T(F^c)$ is hyperinvariant,

$$R(\lambda;T|Y_2)y_2 \in X_T(F^c).$$

Thus, it follows that

$$< x,v > = < R(\lambda;T|Y_2)y_2,u > = 0,$$

and hence $v \in X_T(F^c)^{\perp}$.

Now we can show that v verifies equation

$$(\lambda-T^*)v = u. \tag{12.13}$$

Let again x be arbitrary in X and have a representation (12.12). Then $(\lambda-T)x$ has the representation

$$(\lambda-T)x = (\lambda-T)y_1 + (\lambda-T)y_2, \text{ with } (\lambda-T)y_i \in Y_i, \ i = 1,2.$$

Then, by the definition of v, we have

$$< x,(\lambda-T^*)v > = < (\lambda-T)x,v > = < R(\lambda;T|Y_2)(\lambda-T)y_2,u > =$$

$$= < y_2,u > = < y_1 + y_2,u > = < x,u > .$$

Since x is arbitrary in X, we obtain (12.13).

With this proof of the surjectivity it follows that $\lambda-T^*$ is bijective on $X_T(F^c)^{\perp}$ and hence $\lambda \in \rho[T^*|X_T(F^c)^{\perp}]$ implies (12.11). We remark that we could also choose to refer to Corollary 1.3 instead of referring to Corollary 4.11, at the beginning of the proof.□

12.7. Lemma. Let T *be strongly decomposable and let* $\{G_i\}_1^n$ *be a system of open sets. Then*

$$X_T(\bigcup_{i=1}^n G_i) = \sum_{i=1}^n X_T(G_i).$$

Proof. In view of Proposition 1.5 (i), it suffices to prove

$$X_T(\bigcup_{i=1}^n G_i) \subset \sum_{i=1}^n X_T(G_i).$$

Let $x \in X_T(\bigcup_{i=1}^n G_i)$. Then $Y = X_T[\sigma(x,T)] \in SM(T)$ and by Theorem 12.2, $T|Y$ is decomposable. Since

$$\sigma(T|Y) \subset \sigma(x,T) \subset \bigcup_{i=1}^n G_i,$$

$\{G_i\}_1^n$ is an open cover of $\sigma(T|Y)$. There is a system $\{Y_i\}_1^n \subset SM(T|Y)$ such that

$$Y = \sum_{i=1}^n Y_i \text{ and } \sigma(T|Y_i) \subset G_i,\ i = 1,2,\dots,n.$$

We have

$$Y_i \subset X_T(G_i),\ i = 1,2,\dots,n$$

and since $x \in Y$, it follows that

$$x \in \sum_{i=1}^n Y_i \subset \sum_{i=1}^n X_T(G_i). \ \square$$

The next theorem is not a necessary link in the succession of properties which lead us to the ultimate dual theory. It is, however, interesting at this stage to see some connections between various concepts developed so far.

12.8. Theorem. If T *is strongly decomposable then*

(12.14) $$E(F) = X_T(F^c)^{\perp},\ F \in \mathcal{F}$$

is a 2-spectral capacity possessed by T^*.

Proof. We shall refer to the defining properties of a 2-spectral capacity (Definitions 8.1 and 8.2) as (i), (ii), (iii"), (iv) and (v).

$\mathcal{E}$ as defined by (12.14), clearly satisfies (i); by Proposition 1.8, $\mathcal{E}$ verifies (iv) and by Proposition 12.6 condition (v) holds. We divide the remainder of the proof in parts (A) and (B).

(A). In order to see that $\mathcal{E}$ verifies (ii), let $\{F_n\} \subset \mathcal{F}$ and put

$$F = \bigcap_{n=1}^{\infty} F_n.$$

By Proposition 1.5, for every positive integer N, we have

$$\sum_{n=1}^{N} X_T(F_n^c) \subset X_T(\bigcup_{n=1}^{N} F_n^c) \subset X_T(F^c).$$

Thus, when $N \to \infty$ we obtain

$$\sum_{n=1}^{\infty} X_T(F_n^c) \subset \overline{X_T(F^c)}. \tag{12.15}$$

On the other hand, for every $x \in X_T(F^c)$,

$$\sigma(x,T) \subset F^c = \bigcup_{n=1}^{\infty} F_n^c.$$

$\sigma(x,T)$ being compact, for N sufficiently large,

$$\sigma(x,T) \subset \bigcup_{n=1}^{N} F_n^c,$$

and by Lemma 12.7, we obtain

$$x \in X_T(\bigcup_{n=1}^{N} F_n^c) = \sum_{n=1}^{N} X_T(F_n^c) \subset \sum_{n=1}^{\infty} X_T(F_n^c).$$

Thus (12.15) and the latter inclusions imply

$$X_T(F^c) \subset \sum_{n=1}^{\infty} X_T(F_n^c) \subset \overline{X_T(F^c)}.$$

Therefore

$$\text{c.l.m.}(\bigcup_{n=1}^{\infty} X_T(F_n^c)) = \overline{\sum_{n=1}^{\infty} X_T(F_n^c)} = \overline{X_T(F^c)}. \tag{12.16}$$

Applying the annihilator to (12.16), we obtain successively

$$[\text{c.l.m.}(\bigcup_{n=1}^{\infty} X_T(F_n^c))]^{\perp} = X_T(F^c)^{\perp},$$

$$\bigcap_{n=1}^{\infty} X_T(F_n^c)^{\perp} = X_T(F^c)^{\perp},$$

$$\bigcap_{n=1}^{\infty} E(F_n) = E(F) = E(\bigcap_{n=1}^{\infty} F_n).$$

(B). To prove (iii"), let $\{G_1,G_2\}$ be an open cover of C. The sets

$$F_i = G_i^c, \; i = 1,2$$

are closed and disjoint. By Theorem 3.16,

$$Y = X_T(F_1 \cup F_2) = X_T(F_1) \oplus X_T(F_2),$$

and hence each $x \in Y$ has a unique representation

$$x = x_1 + x_2 \text{ with } x_i \in X_T(F_i), \; i = 1,2.$$

Let $y^* \in X^*$ be arbitrary and define the functional y_1 on Y by

(12.17) $$< x,y_1 > = < x_2,y^* >, \; x \in Y.$$

It clearly follows from (12.17) that y_1 is linear and bounded on Y. By the Hahn-Banach theorem y_1 extends continuously to $y_1^* \in X^*$. We have $y_1^*|X_T(F_1) = 0$ so that

$$y_1^* \in X_T(F_1)^{\perp} \subset X_T(G_1^c)^{\perp} = E(\overline{G}_1).$$

Put

$$y_2^* = y^* - y_1^*.$$

Then for $x_2 \in X_T(F_2)$, with the help of (12.17) we obtain

$$< x_2,y_2^* > = < x_2,y^* > - < x_2,y_1^* > = < x,y_1 > - < x_1+x_2,y_1 > = 0.$$

Hence $y_2^* \in X_T(F_2)^{\perp} \subset E(\overline{G}_2)$ and since

$$y^* = y_1^* + y_2^*$$

is arbitrary, we have

$$X^* = E(\overline{G}_1) + E(\overline{G}_2). \; \Box$$

As an immediate consequence of Theorem 12.8, we note that T^* is 2-decomposable if T is strongly decomposable. In this case, for every $F \in F$, $E(F) = X_{T^*}^*(F)$ and it follows at once that

(12.18) $$X_T(F^c)^{\perp} = X_{T^*}^*(F).$$

Given $T \in B(X)$, *consider the following cases:*

(a) T *is quasidecomposable and its dual* T^* *satisfies property*

(12.19) $$\sigma[T^*|X_T(F^c)^{\perp}] \subset F, \text{ for every } F \in F$$

(b) T *is 2-decomposable.*

12.9. Lemma. In each of cases (a) *and* (b) *above,* T* *has the* 2-SDP.

Proof. Let $\{G_1, G_2\}$ be an open cover of $\sigma(T^*)$. In view of the remarks following Definition 12.5, we can assume that $\{G_1, G_2\}$ covers C. The hypotheses on T allow us to apply part (B) of the proof of Theorem 12.8. This gives us a decomposition of X* into the sum

$$X^* = X_T(\overline{G}_1^{\,c})^{\perp} + X_T(\overline{G}_2^{\,c})^{\perp} .$$

Since $X_T(\overline{G}_i^{\,c})^{\perp}$ is invariant under T* (Proposition 1.8), (12.19) applied to $\overline{G}_1$ and $\overline{G}_2$,

$$\sigma[T^*|X_T(\overline{G}_i^{\,c})^{\perp}] \subset \overline{G}_i, \quad i = 1,2$$

concludes the proof. □

12.10. Lemma. Given T ε B(X), *in each of cases* (a), (b) *above and for every* F ε *F*, $X_T(F^c)^{\perp}$ *is a spectral maximal space of* T*.

Proof. Let Y* ε Inv(T*) satisfy

$$\sigma(T^*|Y^*) \subset \sigma[T^*|X_T(F^c)^{\perp}] \subset F.$$

Let y* ε Y*. By Lemma 12.9 and Theorem 4.9, T* has the SVEP. Then we have

$$\sigma(y^*,T^*) \subset \sigma(T^*|Y^*) \subset F.$$

So for every $x \in X_T(F^c)$, $< x,y^* > = 0$ and hence

$$y^* \in X_T(F^c)^{\perp} .$$

Thus

$$Y^* \subset X_T(F^c)^{\perp}. \square$$

Lemmas 12.9 and 12.10 prove the following

12.11. Theorem. Given T ε B(X), *in each of the cases* (a) *and* (b) *above,* T* *is 2-decomposable on* X*.

12.12. Theorem. Every 2-decomposable operator is quasidecomposable.

Proof. Let $\{G_i\}_1^n$ be an open cover of C. Putting $F_i = G_i^c$, $i = 1,2,\ldots,n$, we have

(12.20) $$\bigcap_{i=1}^{n} F_i = \emptyset.$$

By Theorem 12.11, T* is 2-decomposable and in particular it has the SVEP. Then, for F ε *F*, $X^*_{T^*}(F) \in SM(T^*)$.

By (12.20),

$$X^*_{T^*}(\bigcap_{i=1}^{n} F_i) = X^*_{T^*}(\emptyset) = \{0\}.$$

On the other hand,

$$X^*_{T^*}(\bigcap_{i=1}^{n} F_i) = \bigcap_{i=1}^{n} X^*_{T^*}(F_i)$$

and by (12.18) we obtain

$$X = [\bigcap_{i=1}^{n} X^*_{T^*}(F_i)]^{\perp} = \overline{\sum_{i=1}^{n} \overline{X_T(G_i)}} \subset \overline{\sum_{i=1}^{n} X_T(\overline{G}_i)}.$$

The $X_T(\overline{G}_i)$ are spectral maximal for T and

$$\sigma[T|X_T(\overline{G}_i)] \subset \overline{G}_i, \; i = 1,2,\ldots,n. \; \square$$

12.13. Theorem. T^* *is 2-decomposable on* X^* *iff* T *is 2-decomposable on* X.

Proof. The "if" part was proved by Theorem 12.11. Let T^* be 2-decomposable. Then $\tilde{T} = T^{**}$ is 2-decomposable on $X^{**} = \tilde{X}$. Let

$$J: X \to \tilde{X}$$

be the canonical embedding. T may be identified with $\tilde{T}|JX$. Therefore, T has the SVEP (Proposition 1.10) and the spectral manifolds $X_T(F)$ are defined on $\mathcal{F}$. Now we show that $X_T(F)$ is closed on $\mathcal{F}$. We have

(12.21) $$J[X_T(F)] \subset \tilde{X}_{\tilde{T}}(F),$$

and hence

$$\overline{J[X_T(F)]} = J[\overline{X_T(F)}] \subset \tilde{X}_{\tilde{T}}(F).$$

Then

$$\sigma[T|\overline{X_T(F)}] \subset \sigma[T|\tilde{X}_{\tilde{T}}(F)] \subset F$$

implies the inclusion

$$\overline{X_T(F)} \subset X_T(F)$$

and hence $X_T(F)$ is closed on $\mathcal{F}$.

Our next objective is to show that

(12.22) $$X_T(F) = J^{-1}[\tilde{X}_{\tilde{T}}(F)].$$

By (12.21),

$$X_T(F) \subset J^{-1}[\tilde{X}_{\tilde{T}}(F)],$$

and it follows from

$$T[J^{-1}(\tilde{X}_{\tilde{T}}(F))] = J^{-1}[\tilde{T}\tilde{X}_{\tilde{T}}(F)] \subset J^{-1}[\tilde{X}_{\tilde{T}}(F)] ,$$

that $J^{-1}[\tilde{X}_{\tilde{T}}(F)]$ is invariant under T. Let $\lambda \in F^c$ and let $x \in J^{-1}[\tilde{X}_{\tilde{T}}(F)]$ be arbitrary. Since $\lambda \in \rho[\tilde{T}|\tilde{X}_{\tilde{T}}(F)]$, there is a $y \in \tilde{X}_{\tilde{T}}(F)$ verifying the equation

$$(\lambda-\tilde{T})y = Jx.$$

Then

$$x = J^{-1}(\lambda-\tilde{T})y = (\lambda-T)J^{-1}y$$

and hence λ-T is surjective on $J^{-1}[\tilde{X}_{\tilde{T}}(F)]$. By Corollary 1.3, $\lambda \in \rho(T|J^{-1}[\tilde{X}_{\tilde{T}}(F)])$ and hence

$$\sigma(T|J^{-1}[\tilde{X}_{\tilde{T}}(F)]) \subset F.$$

This implies

$$J^{-1}[\tilde{X}_{\tilde{T}}(F)] \subset X_T(F)$$

thus proving (12.22).

Finally, let $\{G_1,G_2\}$ be an open cover of $\sigma(T) = \sigma(\tilde{T})$. Since $\tilde{T}$ is 2-decomposable, there are $\tilde{Y}_1, \tilde{Y}_2 \in SM(\tilde{T})$ satisfying

$$\tilde{X} = \tilde{Y}_1 + \tilde{Y}_2, \text{ and } \sigma(\tilde{T}|\tilde{Y}_i) \subset G_i, \ i = 1,2.$$

For i = 1,2, denote $F_i = \sigma(\tilde{T}|\tilde{Y}_i)$. We have

(12.23) $$\tilde{X} = \tilde{X}_{\tilde{T}}(F_1) + \tilde{X}_{\tilde{T}}(F_2), \text{ with } F_i \subset G_i, \ i = 1,2.$$

The application of J^{-1} to (12.23), with the help of (12.22) gives us the sought decomposition of X:

$$X = X_T(F_1) + X_T(F_2),$$
$$\sigma[T|X_T(F_i)] \subset F_i \subset G_i, \ i = 1,2. \ \square$$

12.14. *Corollary. Given* $T \in B(X)$, *the following statements are equivalent:*

(i) T *is 2-decomposable;*

(ii) T *is quasidecomposable and*

$$\sigma[T^*|X_T(F^c)^\perp] \subset F, \ F \in \mathcal{F}.$$

Proof. (i) => (ii): follows from Proposition 12.6 and Theorem 12.12.

(ii) => (i): follows from Theorem 12.11 and Theorem 12.13. □

Now we take a closer look at the 2-decomposable operators.

12.15. Theorem. T *is 2-decomposable iff for every open set* $G \subset C$*, there is an invariant subspace* Y *such that*

(12.24) $$\sigma(T|Y) \subset \overline{G} \text{ and } \sigma(T^Y) \cap G = \emptyset.$$

Proof. The "only if" part follows from Corollary 11.11. Let $\{G_1, G_2\}$ be an open cover of $\sigma(T)$. We assume that $\sigma(T) \cap G_1 \cap G_2 \neq \emptyset$, because otherwise G_1 and G_2 disconnect $\sigma(T)$ and the functional calculus provides the 2-decomposability. The cover can be chosen such that neither G_i contains any bounded component of $\rho(T)$. By hypothesis and Proposition 1.17, we can find $Y \in \mathrm{Inv}(T)$ with the properties

$$\sigma(T|Y) \subset \overline{G_1 \cap G_2}, \quad \sigma(T^Y) \cap G_1 \cap G_2 = \emptyset, \quad \sigma(T|Y) \subset \sigma(T).$$

By Proposition 1.15, the latter inclusion implies $\sigma(T^Y) \subset \sigma(T)$. Hence we have

$$\sigma(T^Y) \subset \sigma(T) - (G_1 \cap G_2) = [\sigma(T) - G_1] \cup [\sigma(T) - G_2].$$

That is, $\sigma(T^Y)$ is the disjoint union of two closed sets. Apply the functional calculus to find two subspaces Z_1 and Z_2 of X/Y invariant under T^Y with

(12.25) $$X/Y = Z_1 \oplus Z_2,$$

(12.26) $$\sigma(T^Y|Z_i) \subset \sigma(T) - G_i, \quad i = 1,2.$$

Let $J : X \to X/Y$ be the canonical surjection. Then $J^{-1}Z_i \in \mathrm{Inv}(T)$ and

(12.27) $$X = J^{-1}(X/Y) = J^{-1}(Z_1 \oplus Z_2) = J^{-1}(Z_1) + J^{-1}(Z_2).$$

Next, we prove the inclusion

(12.28) $$\sigma[T|J^{-1}(Z_i)] \subset \sigma(T^Y|Z_i) \cup \sigma(T|Y), \; i = 1,2.$$

Let $\lambda \in \rho(T^Y|Z_i) \cap \rho(T|Y)$ and let $x \in J^{-1}Z_i$ satisfy equation $(\lambda - T)x = 0$. Then

$$(\lambda - T^Y)Jx = 0 \quad \text{with} \quad Jx \in Z_i.$$

Thus $Jx = 0$, $x \in Y$ and hence $x = 0$, by the choice of λ. Now let $x \in J^{-1}Z_i$ be arbitrary. Then $Jx \in Z_i$ and

$$(\lambda - T^Y)Jy = Jx,$$

for some $y \in X$ and $Jy \in Z_i$. Thus, there is $u \in J^{-1}Z_i$ satisfying $(\lambda - T)u = x$. Hence $\lambda - T$ is bijective on $J^{-1}Z_i$ and inclusion (12.28) follows. For $i \neq j$, (12.26) and (12.28) imply the following inclusions

$$\sigma(T|J^{-1}Z_i) \subset [\sigma(T) - G_i] \cup \overline{G_1 \cap G_2} \subset \overline{G}_j,$$

and in view of (12.27), T has the 2-SDP.

Next, we prove that T^Y has the SVEP. Let $f : D \to X$ be analytic and satisfy

$$(\lambda - T)f(\lambda) \in Y \text{ on } D.$$

We may suppose that D is connected. If $D \cap \overline{G}^c \neq \emptyset$ then there is an open disk H such that $H \subset D \cap \overline{G}^c$. Since $H \subset \rho(T|Y)$ by hypothesis, we have

$$f(\lambda) = R(\lambda;T|Y)(\lambda-T)f(\lambda) \in Y \quad \text{on } H,$$

and on all of D by analytic continuation. In case that $D \subset \overline{G}$, D being open we have $D \subset G \subset \rho(T^Y)$. Then $(\lambda-T^Y)Jf(\lambda) = 0$ on D and hence $Jf(\lambda) = 0$. This imples that $f(\lambda) \in Y$ on D and consequently $Y \in AI(T)$. By Theorem 2.11, T^Y has the SVEP.

The subspaces Z_1 and Z_2 of the direct sum (12.25) are spectral maximal for T^Y as ranges of spectral projections. By Theorem 3.9, $Z_1, Z_2 \in AI(T^Y)$. In order to see that the $J^{-1}(Z_i) \in AI(T)$, let $g : D \to X$ be analytic and satisfy conditions

$$(\lambda-T)g(\lambda) \in J^{-1}(Z_i) \quad \text{on } D,\ i = 1,2.$$

We have $(\lambda-T^Y)Jg(\lambda) \in Z_i$ and since $Z_i \in AI(T^Y)$, it follows that $Jg(\lambda) \in Z_i$, or

$$g(\lambda) \in J^{-1}(Z_i) \quad \text{on } D, \quad i = 1,2.$$

Thus, T admits a spectral decomposition in terms of two analytically invariant subspaces. Then Theorem 11.5, confined to $n = 2$, proves that T is 2-decomposable. □

12.16. *Lemma. Let* T *be 2-decomposable and let* $F \subset C$ *be closed. Then for each pair of open sets* $\{G_1, G_2\}$ *which covers* F, *we have*

(12.29) $$X_T(F) \subset X_T(\overline{G}_1) + X_T(\overline{G}_2).$$

Proof. We can avoid the trivial cases by assuming that F, G_1, G_2 intersect $\sigma(T)$ and $\sigma(T) \not\subset G_1 \cup G_2$. Put $K = \overline{G}_1 \cap \overline{G}_2$ and let $Y = X_T(K)$. Then Theorems 3.9 and 2.11 imply that T^Y has the SVEP and by Corollary 11.11, $\sigma(T^Y) \cap K^o = \emptyset$. For notational convenience we put $\hat{X} = X/Y$ and $\hat{T} = T^Y$. Let $x \in X_T(F)$ and let $f : F^c \to X$ be analytic and verify the identity

$$(\lambda-T)f(\lambda) = x \quad \text{on } F^c.$$

Then there is a function $\hat{f} : F^c \to \hat{X}$ analytic on F^c such that for every $\lambda \in F^c$,

(12.30) $$(\lambda-\hat{T})\hat{f}(\lambda) = \hat{x}.$$

Since $\sigma(\hat{T}) \subset (G_1 \cap G_2)^c$, $\hat{f}$ has an analytic extension (with the same notation) to $F^c \cup (G_1 \cap G_2)$ which verifies (12.30). Then $\hat{x} \in \hat{X}_{\hat{T}}(L_1 \cup L_2)$ where $L_1 = F - G_2$ and $L_2 = F - G_1$. Since L_1, L_2 are two disjoint closed sets, we may apply the local functional calculus (1.7) to obtain the decomposition

(12.31) $$\hat{x} = \hat{x}_1 + \hat{x}_2, \text{ with } \hat{x}_i \in \hat{X}_{\hat{T}}(L_i), \quad i = 1,2.$$

Next, we prove that $\hat{x}_i$ may be lifted so that

(12.32) $$x_i \in X_T(L_i \cup K).$$

Then (12.32) follows from (12.31) since we show that

(12.33) $$\hat{X}_{\hat{T}}(L \cup K) \subset X_T(L \cup K)/Y, \quad \text{for any closed } L \subset C.$$

To prove (12.32) let $\hat{y} \in \hat{X}_{\hat{T}}(L \cup K)$ and let $g : (L \cup K)^c \to \hat{X}$ be its local resolvent:

$$\hat{y} = (\lambda-\hat{T})g(\lambda), \quad \lambda \in (L \cup K)^c.$$

Now let $\mu \in (L \cup K)^c$ be fixed and let D_μ be a neighborhood of μ contained in $(L \cup K)^c$. By a part of the proof of Theorem 2.11, there is a disk D with $\mu \in D \subset D_\mu$ and a function $h : D \to X$ analytic with

$$\hat{h}(\lambda) = g(\lambda) \text{ on } D.$$

Then for $\lambda \in D$, $f(\lambda) = (\lambda-T)h(\lambda) - y \in Y$ and hence

$$y = (\lambda-T)[h(\lambda) - R(\lambda;T|Y)f(\lambda)] \quad \text{on } D$$

and on $(L \cup K)^c$ by analytic continuation. This proves that

$$\sigma(y,T) \subset L \cup K$$

and inclusion (12.33) is immediate. Hence (12.32) holds and we can write

(12.34) $$x = x_1 + (x_2 + z), \quad \text{for some } z \in Y.$$

Now (12.29) follows from (12.34). □

12.17. *Lemma. Let* T *be 2-decomposable, let* $F \subset C$ *be closed and let* $\{G_1,G_2\}$ *be a fixed pair of open sets which covers* F. *Then there exists a constant* $M > 0$ *such that for every pair* $\{H_1,H_2\}$ *of open sets with*

(12.35) $$F \subset H_1 \cup H_2 \quad \text{and} \quad \overline{H}_i \subset \overline{G}_i, \quad i = 1,2$$

and for each $x \in X_T(F)$, *there are vectors* x_1, x_2 *satisfying properties:*

(12.36) $$x = x_1 + x_2, \quad \sigma(x_i,T) \subset \overline{H}_i \quad (i=1,2), \quad \|x_1\| + \|x_2\| \le M\|x\|.$$

Proof. Form the direct sum

$$W = X_T(\overline{G}_1) \oplus X_T(\overline{G}_2)$$

and consider the continuous linear mapping $k : W \to X$ defined by

$$k(x_1 \oplus x_2) = x_1 + x_2, \quad \sigma(x_i,T) \subset \overline{G}_i, \quad i = 1,2.$$

By Lemma 12.16, the range of k contains $X_T(F)$. Hence $W_0 = k^{-1}[X_T(F)]$ is a closed subspace of W and the restriction $k_0 = k|W_0$ is surjective on $X_T(F)$. The closed graph theorem gives an $M > 0$ such that (12.36) holds for $H_i = G_i$.

Now let $\{H_1,H_2\}$ be an arbitrary pair of open sets satisfying (12.35). Then

$$W_1 = X_T(\overline{H}_1) \oplus X_T(\overline{H}_2) \subset W$$

is a closed subspace of W. Let $k_1 = k|W_1$. Since $k_1|W_0 = k_0$ (because the range of k_1 also contains $X_T(F)$ by Lemma 12.16) the conclusions (12.36) also apply to $\{H_1,H_2\}$.

12.18. Theorem. If $T|Y$ *is 2-decomposable for every* $Y \in SM(T)$ *then* T *is strongly decomposable.*

Proof. In view of Theorem 12.2, it suffices to prove that T is decomposable. Then that proof may be applied to $T|Y$. Now Lemma 12.7 applies so that for any open cover $\{G_i\}_1^n$ of $\sigma(T)$, we have

$$X = X_T(\bigcup_{i=1}^n G_i) = \sum_{i=1}^n X_T(G_i) \subset \sum_{i=1}^n X_T(\overline{G}_i) \subset X. \quad \square$$

12.19. Theorem. Let T *be 2-decomposable. If* $W \in SM(T^*)$ *then* $T^*|W$ *is 2-decomposable.*

Proof. By Theorem 12.13 T^* is 2-decomposable, hence we can use for W the representation

$$W = X^*_{T^*}(F), \quad \text{where} \quad F = \sigma(T^*|W).$$

Let $\{G_1, G_2\}$ be an arbitrary but fixed pair of open sets covering F. We shall prove that there are two closed sets F_1, F_2 such that $F_i \subset G_i$, $i = 1,2$ and

$$W = X^*_{T^*}(F_1) + X^*_{T^*}(F_2) \tag{12.37}$$

from which it will follow that $T^*|W$ is 2-decomposable.

Let $u \in W$ be an arbitrary unit vector. Let A be the family of all pairs of open sets $\{H_1, H_2\}$ satisfying (12.35). Then A forms a directed set under inclusion. For notational purposes index A by $\alpha \in A$. Now let $M > 0$ be the constant determined in Lemma 12.17. By Lemma 12.17 applied to T^*, for each $\alpha \in A$ there are vectors u_i^α with the following properties:

$$u = u_1^\alpha + u_2^\alpha, \quad \sigma(u_i^\alpha, T^*) \subset \overline{H}_{i,\alpha} \quad (i=1,2) \quad \text{and} \quad \|u_1^\alpha\| + \|u_2^\alpha\| \le M.$$

Hence $\{u_i^\alpha : \alpha \in A\}$, i=1,2 are two bounded nets in X^* and by Alaoglu's theorem, each has a subnet (denoted without changing index) converging to u_1, u_2, respectively. For fixed $\beta \in A$ it is clear that if $\alpha \ge \beta$ then

$$\sigma(u_i^\alpha, T^*) \subset \overline{H}_{i,\beta}, \quad i = 1,2. \tag{12.38}$$

By Lemma 12.10, every spectral maximal space of T^* is the annihilator of a linear manifold in X, hence every subspace $X^*_{T^*}(\overline{H}_{i,\alpha})$ is weak*-closed. Thus

$$u_i \in X^*_{T^*}(\overline{H}_{i,\alpha}), \quad i = 1,2; \quad \alpha \in A,$$

and moreover,

$$u_i \in \bigcap_{\alpha \in A} X^*_{T^*}(\overline{H}_{i,\alpha}), \; i = 1,2.$$

Let $F_i = \bigcap_{\alpha \in A} \overline{H}_{i,\alpha}$, $i = 1,2$. Clearly, $F_1 \cup F_2 = F$ and $u = u_1 + u_2$. Hence

$$u \in X^*_{T^*}(F_1) + X^*_{T^*}(F_2) \subset W$$

and since u was arbitrary in W, we obtain (12.37). □

12.20. Corollary. Let T *be 2-decomposable. Then* T *is strongly decomposable.*

Proof. By Theorem 12.18 it suffices to prove that $T|Y$ is 2-decomposable for every $Y \in SM(T)$. Let $S = T^{**}$. Since T^* is 2-decomposable by Theorem 12.13, it follows from Theorem 12.19 that $S|W$ is 2-decomposable for every $W \in SM(S)$. Now let F be closed and put $V = X^{**}$.

For any open cover $\{G_1, G_2\}$ of F we have by Theorem 12.19

$$V_S(F) = V_S(F_1) + V_S(F_2) \tag{12.39}$$

for closed sets $F_1, F_2 \subset F$. As in the proof of Theorem 12.13, with $k : X \to V$ as the canonical embedding, we can write with the help of (12.39)

$$X_T(F) = k^{-1}[V_S(F)] =$$

$$= k^{-1}[V_S(F_1)] + k^{-1}[V_S(F_2)] = X_T(F_1) + X_T(F_2).$$

This completes the proof. □

We now summarize these results in the following

12.21. Theorem. For $T \in B(X)$, *the following statements are equivalent:*

(*i*) T *is decomposable.*

(*ii*) T *is 2-decomposable.*

(*iii*) T *is strongly decomposable.*

(*iv*) $T|Y$ *is decomposable for* $Y \in SM(T)$.

(*v*) T^Y *is decomposable for* $Y \in SM(T)$.

(*vi*) $T|Y$ *is decomposable for* $Y = \overline{X_T(G)}$, G *open.*

(*vii*) T^* *is decomposable.*

Proof. The equivalence of (i), (ii), (iii), (iv), (v) and (vii) follows from the foregoing properties. The implication (vi) => (i) follows by taking $G = \mathbb{C}$. We prove (v) => (vi): Let G be open, $Y = \overline{X_T(G)}$ and put $Z = Y^\perp$. Then $Z \in SM(T^*)$ by Lemma 12.10 and $(T^*)^Z$ is decomposable by (v). But $(T^*)^Z$ can be identified with $(T|Y)^*$ and hence (vi) follows from the equivalence (i) <=> (vii). □

As a by-product, we obtain an example of an analytically invariant subspace which is not necessarily spectral maximal (see Appendix A.1.).

12.22. *Corollary. For* $T \in D(X)$ *and every open* $G \subset C$, $\overline{X_T(G)} \in AI(T)$.

Proof. By Theorem 12.21, T^Y is decomposable whence $Y = \overline{X_T(G)}$. Then T^Y has the SVEP and hence by Theorem 2.11, $Y \in AI(T)$. □

§ 13. *Spectral decompositions of unbounded operators.*

If we restrict the invariant subspaces to the domain D_T of a given closed linear operator T then the extension of analytically invariant and spectral maximal spaces to the unbounded case is straightforward. Consequently, the concept of weakly decomposable operator extends to the unbounded case. We shall write F and K for the families of closed and compact sets in C, respectively.

13.1. *Definition. A strong spectral capacity in* X *is an application*

$$E : F \to S(X)$$

that satisfies the following conditions:

(I) $E(\emptyset) = \{0\}$, $E(C) = X$;

(II) $E(\bigcap_{n=1}^{\infty} F_n) = \bigcap_{n=1}^{\infty} E(F_n)$, *for every sequence* $\{F_n\} \subset F$;

(III) *For every* $F \in F$ *and every open cover* $\{G_i\}_1^n$ *of* F,

$$E(F) = \sum_{i=1}^{n} E(F \cap \overline{G}_i);$$

(IV) *For every* $F \in F$, *the linear manifold*

$$E_0(F) = \{x \in E(K) : K \in K \text{ and } K \subset F\}$$

is dense in $E(F)$.

For $F = C$, (III) becomes

(III')
$$X = \sum_{i=1}^{n} E(\overline{G}_i),$$

where $\{G_i\}_1^n$ is an open cover of C. Conditions (I), (II) and (III') define the original concept of spectral capacity as given by Definition 8.1.

In the special case, $F = \mathbb{C}$, condition (IV) asserts that

(IV') $$E_0(\mathbb{C}) = \{x \in E(K): K \in \mathcal{K}\}$$

is dense in X.

Condition (IV) is equivalent to

(IV'') For every $F \in \mathcal{F}$ there exists a nondecreasing sequence $\{K_n\} \subset \mathcal{K}$, such that

$$E(F) = \overline{\bigcup_{n=1}^{\infty} E(K_n)}.$$

13.2. Definition. A closed linear operator $T:D_T(\subset X) \to X$ *is said to possess a strong spectral capacity* E *if it has nonvoid resolvent set and satisfies conditions:*

(V) $E(K) \subset D_T$, *for all* $K \in \mathcal{K}$;

(VI) $T[E(F) \cap D_T] \subset E(F)$, *for all* $F \in \mathcal{F}$;

(VII) *The restriction* $T_F = T|E(F) \cap D_T$ *has the spectrum*

$$\sigma(T_F) \subset F, \text{ for each } F \in \mathcal{F}.$$

Remark. It follows from (IV') and (V) that every T with a strong spectral capacity is densely defined in X.

13.3. Theorem. If T *possesses a strong spectral capacity* E *then for every* $K \in \mathcal{K}$, *the restriction* $T_K = T|E(K)$ *is a bounded decomposable operator on* E(K) *possessing the spectral capacity* E_K *defined by*

(13.1) $$E_K(F) = E(K \cap F), \text{ for all } F \in \mathcal{F}.$$

Proof. T being closed, T_K is closed and defined everywhere on $Y = E(K)$. By the closed graph theorem, T_K is bounded. The application

$$E_K:\mathcal{F} \to S(Y),$$

defined by (13.1) gives

(13.2) $$E_K(F) = E(K) \cap E(F) = Y \cap E(F), \text{ for all } F \in \mathcal{F}.$$

The proof proceeds by showing that E_K satisfies conditions (i), (ii), (iii), (iv) and (v) of Definitions 8.1 and 8.2. Conditions (i) and (ii) are immediate consequences of (13.2), (I) and (II). Let $\{G_i\}_1^n$ be an open cover of K. For F = K, (III) becomes

$$Y = \sum_{i=1}^{n} E(K \cap \overline{G}_i) = \sum_{i=1}^{n} E_K(\overline{G}_i)$$

and hence it follows that E_K is a spectral capacity on Y. Moreover, with the help of (V) and (VI) we obtain

$$T_K[E_K(F)] = T[E(K) \cap E(F)] = T[E(K \cap F)] \subset E(K \cap F) = E_K(F), \quad F \in F.$$

Consequently, E_K verifies condition (iv) of Definition 8.2. Finally, (V) and (VII) lead us to (v) of Definition 8.2 as follows:

$$\sigma[T_K | E_K(F)] = \sigma[T|E(K \cap F)] \subset K \cap F \subset F, \quad F \in F.$$

Now Theorem 11.9 concludes the proof. □

13.4. Theorem. Every T *with a strong spectral capacity has the* SVEP.

Proof. Let $f : G \to D_T$ be analytic and verify equation

(13.3) $$(\lambda-T)f(\lambda) = 0 \quad \text{on an open} \quad G \subset C.$$

There is no loss of generality in assuming that G is relatively compact. Let $\{G_1, G_2\}$ be an open cover of C with $G_1 (\supset \overline{G})$ relatively compact and $\overline{G}_2 \cap \overline{G} = \emptyset$. The strong spectral capacity E of T provides the following decomposition of X:

$$X = E(\overline{G}_1) + E(\overline{G}_2).$$

In view of Lemma 4.8, for every $\lambda \in G$ there is a neighborhood $H(\subset G)$ of λ and there are analytic functions $f_i : H \to E(\overline{G}_i)$, $i = 1,2$ such that

(13.4) $$f(\mu) = f_1(\mu) + f_2(\mu), \quad \text{for all } \mu \in H.$$

Since the ranges of both f and f_1 are contained in D_T, we have

$$f_2(H) \subset E(\overline{G}_2) \cap D_T.$$

Equation (13.3) written as

$$(\mu-T)[f_1(\mu) + f_2(\mu)] = 0$$

gives rise to

$$(\mu-T)f_1(\mu) = (T-\mu)f_2(\mu) = g(\mu) \in E(\overline{G}_1) \cap E(\overline{G}_2) \text{ on } H.$$

With the help of (VII),

$$\sigma[T|E(\overline{G}_1 \cap \overline{G}_2)] \subset \overline{G}_1 \cap \overline{G}_2,$$

an analytic function $h : H \to E(\overline{G}_1) \cap E(\overline{G}_2)$ is defined as follows:

$$h(\mu) = [T|E(\overline{G}_1 \cap \overline{G}_2) - \mu]^{-1} g(\mu) \in E(\overline{G}_1) \cap E(\overline{G}_2) \text{ on } H.$$

We have

$$h(\mu) - f_2(\mu) \in E(\overline{G}_2) \cap D_T, \text{ for all } \mu \in H.$$

The definitions of the functions h and g produce the following identity on H:

(13.5) $$[T|E(\overline{G}_2) \cap D_T - \mu][h(\mu) - f_2(\mu)] = 0.$$

By (VII)

$$\sigma[T|E(\overline{G}_2) \cap D_T] \subset \overline{G}_2$$

and therefore $\mu \in H \subset \rho[T|E(\overline{G}_2) \cap D_T]$. Consequently, (13.5) implies

$$f_2(\mu) = h(\mu) \in E(\overline{G}_1)$$

and then (13.4) implies that $f(\mu) \in E(\overline{G}_1)$ on H. Since by Theorem 13.3, $T|E(\overline{G}_1)$ is decomposable, it has the SVEP and hence (13.3) implies that

$$f(\mu) = 0 \quad \text{on } H,$$

and on all of G, by analytic continuation. □

13.5 *Theorem. Given* T *with the strong spectral capacity* E, *for every* $K \in \mathcal{K}$, $E(K)$ *is a spectral maximal space of* T.

Proof. (A). Let $Y(\subset D_T)$ be a subspace of X invariant under T such that

$$\sigma(T|Y) \subset \sigma[T|E(K)],$$

and let $x \in Y$ be arbitrary. The proof will be brought to its conclusion by showing that $x \in E(K)$. By (VII),

$$\sigma[T|E(K)] \subset K.$$

(B). Thus the hypotheses are : $Y(\subset D_T)$ is invariant under T, $x \in Y$ and

$$\sigma(T|Y) \subset K. \tag{13.6}$$

Let $\{G_1, G_2\}$ be an open cover of C with $K \subset G_1$, G_1 relatively compact and $\overline{G}_2 \cap K = \emptyset$. By (III),

$$x = x_1 + x_2 \quad \text{with} \quad x_i \in E(\overline{G}_i), \quad i = 1,2.$$

Note that since $x \in Y \subset D_T$ and $x_1 \in E(\overline{G}_1) \subset D_T$, it follows that $x_2 = x - x_1 \in D_T$. In view of (13.6), $\tilde{x}$ is defined on K^c and verifies equation

$$(\lambda - T)\tilde{x}(\lambda) = x, \quad \text{for all} \quad \lambda \in K^c. \tag{13.7}$$

Furthermore, since by (VII)

$$\sigma[T|E(\overline{G}_2) \cap D_T] \subset \overline{G}_2, \tag{13.8}$$

there is a function

$$\tilde{x}_2 : \overline{G}_2{}^c \to E(\overline{G}_2) \cap D_T$$

analytic and verifying equation

$$(\lambda - T)\tilde{x}_2(\lambda) = x_2 \quad \text{on } \overline{G}_2{}^c. \tag{13.9}$$

Combining (13.7) and (13.9), we obtain

$$(\lambda-T)[\tilde{x}(\lambda)-\tilde{x}_2(\lambda)] = x-x_2 = x_1, \text{ for all } \lambda \in G = \overline{G}_2^c \cap K^c,$$

and $\tilde{x} - \tilde{x}_2$ is analytic on G. Since by Theorem 13.4, T has the SVEP, $\tilde{x}_1$ analytic and verifying equation

$$(\lambda-T)\tilde{x}_1(\lambda) = x_1 \text{ on } G, \tag{13.10}$$

is uniquely determined by

$$\tilde{x}_1(\lambda) = \tilde{x}(\lambda)-\tilde{x}_2(\lambda), \text{ for all } \lambda \in G. \tag{13.11}$$

If Γ is an admissible contour surrounding K and contained in G, in view of (13.8) we have

$$\int_\Gamma \tilde{x}_2(\lambda)d\lambda = 0. \tag{13.12}$$

With the help of (13.6), (13.11) and (13.12) we obtain

$$x = \frac{1}{2\pi i}\int_\Gamma R(\lambda;T|Y)x d\lambda = \frac{1}{2\pi i}\int_\Gamma \tilde{x}(\lambda)d\lambda = \frac{1}{2\pi i}\int_\Gamma \tilde{x}_1(\lambda)d\lambda .$$

Therefore, to prove that $x \in E(\overline{G}_1)$, it suffices to show that $\tilde{x}_1(\lambda) \in E(\overline{G}_1)$ on Γ. Apply Lemma 4.8 to the function $\tilde{x}_1 : G \to D_T$ analytic on G. For each $\lambda \in G$, there is a neighborhood $H \subset G$ of λ and there are analytic functions

$$g_i : H \to E(\overline{G}_i), \ i = 1,2 \tag{13.13}$$

such that

$$\tilde{x}_1(\mu) = g_1(\mu) + g_2(\mu), \text{ for all } \mu \in H. \tag{13.14}$$

Since $\tilde{x}_1(\mu) \in D_T$ and $g_1(\mu) \in E(\overline{G}_1) \subset D_T$, it follows from (13.14) that

$$g_2(\mu) \in E(\overline{G}_2) \cap D_T, \text{ for all } \mu \in H.$$

Relations (13.10) and (13.14) imply that

$$(\mu-T)[g_1(\mu) + g_2(\mu)] = x_1 \text{ on } H$$

and hence

$$x_1-(\mu-T)g_1(\mu) = (\mu-T)g_2(\mu).$$

So

$$f(\mu) = (\mu-T)g_2(\mu) \in E(\overline{G}_1) \cap E(\overline{G}_2) = E(\overline{G}_1 \cap \overline{G}_2), \text{ on } H.$$

The analytic function

$$h(\mu) = [\mu-T|E(\overline{G}_1\cap\overline{G}_2)]^{-1}f(\mu) \in E(\overline{G}_1 \cap \overline{G}_2), \text{ on } H$$

gives rise to

$$(\mu-T)[h(\mu)-g_2(\mu)] = 0 \text{ on } H.$$

By the SVEP of T,

$$g_2(\mu) = h(\mu) \in E(\overline{G}_1 \cap \overline{G}_2) \subset E(\overline{G}_1) \quad \text{on } H$$

and hence

$$\tilde{x}_1(\mu) = g_1(\mu) + g_2(\mu) \in E(\overline{G}_1).$$

Since λ is arbitrary on G and hence on Γ, we have

$$\tilde{x}_1(\lambda) \in E(\overline{G}_1) \quad \text{on } \Gamma.$$

Thus, $x \in E(\overline{G}_1)$ for every relatively compact open $G_1 \supset K$. So, if $\{G_{1n}\}$ is the sequence of the n^{-1} -neighborhoods of K, then

$$x \in \bigcap_{n=1}^{\infty} E(\overline{G}_{1n}) = E(\bigcap_{n=1}^{\infty} \overline{G}_{1n}) = E(K). \square$$

A slightly different version of the foregoing proof can show that if $G \subset \mathbb{C}$ is open such that $E(\overline{G}) \subset D_T$, then $E(\overline{G})$ is an analytically invariant subspace under T (Erdelyi [3]).

13.6. *Theorem. If* T *possesses a strong spectral capacity* E *then* E *is unique.*

Proof. Suppose that T has a second strong spectral capacity E'. Let $K \in \mathcal{K}$ be arbitrary and let $x \in E'(K)$. By (VII),

$$\sigma[T|E'(K)] \subset K.$$

Denote $Y = E'(K)$ and repeat part (B) of the proof of Theorem 13.5 to obtain $x \in E(K)$. Thus, $E'(K) \subset E(K)$. By symmetry, $E(K) \subset E'(K)$ and consequently

$$E(K) = E'(K), \quad \text{for all } K \in \mathcal{K}. \tag{13.15}$$

Now, let $F \in \mathcal{F}$ be arbitrary. By (IV") there exists a nondecreasing sequence $\{K_n\} \subset \mathcal{K}$ such that for every strong spectral capacity E,

$$E(F) = \overline{\bigcup_{n=1}^{\infty} E(K_n)}.$$

Then, with the help of (13.15), we have

$$E'(F) = \overline{\bigcup_{n=1}^{\infty} E'(K_n)} = \overline{\bigcup_{n=1}^{\infty} E(K_n)} = E(F). \square$$

13.7. *Theorem. If* T *possesses a strong spectral capacity and* $\infty \in \rho(T)$, *then* T *is weakly decomposable.*

Proof. Let $\{G_i\}_1^n$ be an open cover of $\sigma(T)$. Since, by hypothesis $\sigma(T)$ is compact, there is a relatively compact open neighborhood H of $\sigma(T)$. The sets

$$H_i = H \cap G_i, \ i = 1,2,\ldots,n$$

form a relatively compact open cover of $\sigma(T)$. Let H_0 be an open set such that $\{H_i\}_0^n$ covers $\mathbb{C}$ and $\overline{H}_0 \cap \sigma(T) = \emptyset$. Let $\mathcal{E}$ be the strong spectral capacity possessed by T. By (VII) and with the help of Corollary 8.5 applicable to $\mathcal{E}$, we have

$$\sigma[T|\mathcal{E}(\overline{H}_0) \cap D_T] \subset \overline{H}_0 \cap \sigma(T) = \emptyset$$

and hence

$$\mathcal{E}(\overline{H}_0) \cap D_T = \{0\}. \tag{13.16}$$

By (III') we have

$$X = \sum_{i=0}^{n} \mathcal{E}(\overline{H}_i). \tag{13.17}$$

Relations (13.16) (13.17) imply

$$D_T \subset \sum_{i=1}^{n} \mathcal{E}(\overline{H}_i),$$

and since D_T is dense in X, we have

$$X = \overline{\sum_{i=1}^{n} \mathcal{E}(\overline{H}_i)}.$$

Furthermore, (VII) implies

$$\sigma[T|\mathcal{E}(\overline{H}_i)] \subset \overline{H}_i \subset \overline{G}_i, \ i = 1,2,\ldots,n.$$

By Theorem 13.5, for every i, $\mathcal{E}(\overline{H}_i)$ is spectral maximal for T. □

13.8. *Theorem. Let* T *have the strong spectral capacity* $\mathcal{E}$. *For every* $x \in X$, *there exists a nonvoid open set* G *and a sequence* $\{f_n\}$ *of functions analytic on* G *with values in* D_T *such that* $(\lambda - T)f_n(\lambda)$ *converges to* x *uniformly on every compact subset of* G.

Proof. Let $x \in X$ be arbitrary but fixed. Choose $F \in \mathcal{F}$ such that $x \in \mathcal{E}(F)$ and $F \neq \mathbb{C}$. By (IV'') there exists a nondecreasing sequence $\{K_m\} \subset \mathcal{K}$ such that

$$\mathcal{E}(F) = \overline{\bigcup_{m=1}^{\infty} \mathcal{E}(K_m)}.$$

There is a sequence $\{x_n\} \subset \bigcup_{m=1}^{\infty} E(K_m)$ such that $x_n \to x$. For every n, $x_n \in E(K_m)$ (hence $x_n \in D_T$), for some m. Note that

$$\bigcap_{m=1}^{\infty} K_m^c \supset F^c = G.$$

For every n, there exists a function $f_n : G \to D_T$ analytic on G such that

$$(\lambda - T) f_n(\lambda) = x_n, \text{ for all } \lambda \in G.$$

Thus, on every compact subset of G, we have

$$\lim_{n \to \infty} (\lambda - T) f_n(\lambda) = \lim_{n \to \infty} x_n = x. \square$$

NOTES AND COMMENTS.

The concept of decomposable operator was introduced by Foiaş [2]. The original paper contains the proofs of Theorem 11.2 and of properties (11.a), (11.b), (11.d), (11.e). A forerunner of some basic concepts developed for decomposable operators (Foiaş [1]) set the foundations of the "generalized scalar operators", which together with the "generalized spectral operators" (Colojoară [1], Maeda [1]) are defined by means of a spectral distribution (i.e. a L. Schwartz-type multiplicative vector distribution). The theory of the generalized spectral operators, systematically presented in Colojoară and Foiaş [3] offers to the reader an important application of decomposable operators. Chapter 2 of the last reference is dedicated to the general theory of decomposable operators. The theoretical foundations of decomposable operators as well as of the generalized spectral operators form the subject of Colojoară [2].

Some extensive studies on the functional calculus of decomposable operators including direct proofs of Corollaries 11.6, 11.7 and of Theorems 11.13, 11.14 form the topic of Colojoară and Foiaş [2]. As indicated in some remarks following Theorem 11.14, some deep results on this functional calculus were obtained by Apostol [5,6].

Some of the herein contained characterizations of decomposable operators have the following sources: Theorem 11.5, Lange [2]; Theorem 11.8, Apostol [5]; Theorem 11.9, Foiaş [3]. Proposition 11.10 was proved by Apostol [3]. Corollary 11.11 was given in a restricted formulation by Jafarian and Vasilescu [1] .

The duality theory of spectral decompositions has been developed as the result of a relatively small number of papers. Definition 12.1, Theorem 12.2, Lemma 12.3 and Theorem 12.4 are due to Apostol [3]; Definition 12.5, Lemma 12.7 and Theorem 12.18 appeared in Plafker [1]; Proposition 12.6, parts of Lemmas 12.9, 12.10 and Theorem 12.11 as well as Theorem 12.12 were proved by Frunză [1].

Theorem 12.13 and Corollary 12.14 were proved by Lange [4] as well as Theorem 12.15, Theorem 12.19, Corollary 12.20 and Theorem 12.21 are among the results obtained by Lange [3].

An important result expressed by Lemma 12.16 is due to Radjabalipour [2] and private communication. By using the concept of "almost localized spectrum" introduced by Vasilescu [4] and some related results of Frunză [3,4], Radjabalipour obtained an independent proof of the equivalence (i) <=> (ii) of Theorem 12.21. Lemma 12.16 was instrumental for our proof of Theorem 12.21. The first part of the proof of Lemma 12.17 is due to Vasilescu [4].

Section 13 on unbounded operators is based on Erdelyi [1,3].

This chapter obviously does not cover all known properties of decomposable operators. Just to mention a few interesting properties which are beyond the scope of this work we refer to three additional papers. An illuminating characterization of spectral maximal spaces when they belong to decomposable operators was given by Foiaş [4]. He proved that these subspaces coincide with both strong and weak spectral manifolds introduced by Bishop [1]. As an interesting analogy with the Dunford-type spectral operators, Foiaş [5] introduced the scalar part of decomposable operators. Finally, it is worth mentioning that Apostol [4] developed a topology in which the class of decomposable operators is closed.

APPENDIX

A.1. *An example of an analytically invariant subspace which is not absorbent.*

A.1.1. Preliminaries. Let $X = C[a,b]$ be the Banach space of complex-valued continuous functions on $[a,b]$ endowed with the norm

$$\| x \| = \sup_{t \in [a,b]} |x(t)|, \quad x \in X.$$

Define the multiplication operator $T \in B(X)$ as follows

$$Tx(t) = tx(t), \quad t \in [a,b].$$

For every complex $\lambda \notin [a,b]$, $R(\lambda;T)$ is defined by

$$R(\lambda;T)x(t) = (\lambda-t)^{-1}x(t), \text{ for all } x \in X.$$

Hence, $\sigma(T) \subset [a,b]$. Conversely, for $\lambda \in \rho(T)$, let

$$x = R(\lambda;T)u, \text{ where } u(t) \equiv 1.$$

Then

$$(\lambda-t)x(t) = 1$$

and hence $\lambda \notin [a,b]$ thus implying that $[a,b] \subset \sigma(T)$. Consequently,

(A.1.1) $$\sigma(T) = [a,b].$$

Now, let $F \subset [a,b]$ be closed and define

$$Y_F = \{x \in X: \text{supp } x \subset F\}.$$

Clearly, $Y_F \in \text{Inv}(T)$. We will show that

(A.1.2) $$\sigma(T|Y_F) \subset F.$$

This will follow from the fact that for $\lambda \in F^c$, $\lambda-T|Y_F$ is bijective. Let $\lambda \in F^c$ and let $x \in Y_F$ satisfy

$$(\lambda-T)x = 0.$$

Then

$$(\lambda-t)x(t) = 0, \text{ for all } t \in [a,b].$$

Therefore,

$$x(t) = 0 \text{ for all } t \in F$$

and hence $x = 0$. For the proof of the surjectivity, let $x \in Y_F$ be arbitrary and define $y \in C[a,b]$ by

$$y(t) = \begin{cases} (\lambda-t)^{-1}x(t), & \text{if } t \in F; \\ 0, & \text{if } t \in F^c. \end{cases}$$

Since supp $x \subset F$, we have $y \in Y_F$. Equation

$$(\lambda-T)y = x$$

proves that $\lambda-T$ is surjective on Y_F. Thus $\lambda-T$ is bijective on Y_F and property (A.1.2) follows.

During our next step, we show that $Y_F \in SM(T)$. Let $Z \in Inv(T)$ satisfy

$$\sigma(T|Z) \subset \sigma(T|Y_F) \subset F,$$

and let $x \in Z$. Fix $\lambda \in [a,b] \cap F^c$. Since $\lambda \in \rho(T|Z)$, $\lambda-T|Z$ is surjective and hence there exists a function $y \in Z$ such that

$$(\lambda-T)y = x,$$

i.e.
$$(\lambda-t)y(t) = x(t), \text{ for all } t \in [a,b] .$$

In particular for $t = \lambda$, we have $x(\lambda) = 0$ which proves that supp $x \subset F$ and hence $Z \subset Y_F$.

We proceed by proving that $T \in D(X)$. Let $\{G_i\}_1^n$ be an open cover of $\sigma(T)$. Let $\{\alpha_i\}_1^n$ be a continuous partition of the unity subordinate to the cover $\{G_i\}$, i.e.

$$\alpha_i \in C[a,b], \text{ supp } \alpha_i \subset G_i, \; i = 1,2,\ldots,n$$

$$\sum_{i=1}^{n} \alpha_i(t) = 1, \text{ for all } t \in [a,b].$$

Denote $F_i = \text{supp } \alpha_i$, $1 \leq i \leq n$. Then for every i, $Y_{F_i} \in SM(T)$ and for every $x \in X$, we have

$$x = \sum_{i=1}^{n} \alpha_i x, \text{ with } \alpha_i x \in Y_{F_i}.$$

Thus, we have

$$X = \sum_{i=1}^{n} Y_{F_i},$$

and by (A.1.2),

$$\sigma(T|Y_{F_i}) \subset \text{supp } \alpha_i \subset G_i, \; i = 1,2,\ldots,n.$$

Finally, we show that for every $x \in X$,

$$\sigma(x,T) = \operatorname{supp} x. \tag{A.1.3}$$

By (A.1.2) we have

$$\sigma(x,T) \subset \sigma(T|Y_{\operatorname{supp} x}) \subset \operatorname{supp} x.$$

Conversely,

$$[(\lambda-T)\tilde{x}(\lambda)](t) = x(t), \text{ for every } \lambda \in \rho(x,T) \text{ and } t \in [a,b].$$

For $\lambda \in \rho(x,T) \cap [a,b]$, we have

$$\tilde{x}(\lambda) = 0,$$

and hence x vanishes on $\rho(x,T) \cap [a,b]$. This, however, implies that

$$\operatorname{supp} x \subset \sigma(x,T),$$

thus proving (A.1.3).

A.1.2. *The Example.* Let $X = [-1,1]$ and consider the multiplication operator T defined by

$$Tx(t) = tx(t), \ t \in [-1,1].$$

Let

$$G = (-2,0) \cup (0,2).$$

By Corollary 12.22, $Y = \overline{X_T(G)} \in AI(T)$. In view of (A.1.3),

$$X_T(G) = \{x \in X: \operatorname{supp} x \subset G\}.$$

The sequence $\{x_n\} \subset X_T(G)$, defined for every n by

$$x_n(t) = \begin{cases} t^3 - t + \frac{1}{n^3} - \frac{1}{n}, & \text{for } -1 \le t \le -\frac{1}{n}; \\ 0, \text{ for } & -\frac{1}{n} < t < \frac{1}{n}; \\ t^3 - t - \frac{1}{n^3} + \frac{1}{n}, & \frac{1}{n} \le t \le 1 \end{cases}$$

converges uniformly to a function $x \in X$ defined by

$$x(t) = t^3 - t, \ t \in [-1,1].$$

The limit function $x \in \overline{X_T(G)}$. We clearly have

$$\sigma(x,T) = \operatorname{supp} x = [-1,1]$$

and then

$$\sigma(T|Y) = [-1,1] \subset \overline{G}.$$

Y is not absorbent because for $0 \in \sigma(T|Y)$, the equation

$$Ty = x$$

has the solution

$$y(t) = t^2 - 1$$

and it is easily seen that $y \notin Y$.

In view of Theorem 3.7, Y is neither spectral maximal.

A.2. *The set-spectra of decomposable operators.*

A.2.1. *Definition. Given* $T \in D(X)$, *a closed subset* F *of* $\sigma(T)$ *is called a set-spectrum of* T *if there exists* $Y \in SM(T)$ *such that*

$$\sigma(T|Y) = F$$

or, equivalently, if

$$\sigma[T|X_T(F)] = F.$$

A.2.2. *Theorem. Given* $T \in D(X)$, *for every closed subset* F *of* $\sigma(T)$, *there is a largest set-spectrum* F_0 *contained in* F. F_0 *has the following properties*

(A.2.1) $$F_0 = \sigma[T|X_T(F)],$$

(A.2.2) $$X_T(F_0) = X_T(F).$$

Proof. F_0 as defined by (A.2.1) is a set-spectrum contained in F. Let $\sigma(T|Y)$ be any set-spectrum contained in F. Our objective is to show that $\sigma(T|Y) \subset F_0$. The hypothesis on Y implies

$$Y = X_T[\sigma(T|Y)] \subset X_T(F),$$

and hence we have

$$\sigma(T|Y) \subset \sigma[T|X_T(F)] = F_0.$$

In order to obtain (A.2.2), it suffices to prove the inclusion

$$X_T(F) \subset X_T(F_0).$$

Let $x \in X_T(F)$ be arbitrary. Then we have

$$\sigma(x,T) \subset \sigma[T|X_T(F)] = F_0$$

and hence $x \in X_T(F_0)$. □

A.2.3. *Theorem.* *Given* $T \in D(X)$, *let* $F \subset \sigma(T)$ *be closed and denote by* F^I *the interior of* F *in the topology of* $\sigma(T)$. *Then for the largest set-spectrum* F_0 *contained in* F *we have*

$$\overline{F^I} \subset F_0 \subset F. \tag{A.2.3}$$

Proof. Let $\lambda \in F^I$. For every neighborhood $V(\lambda)$ of λ open in the topology of $\sigma(T)$, the set $V(\lambda) \cap F^I$ is a nonvoid set open in $\sigma(T)$. By Proposition 7.2, there exists $Y \in SM(T)$ and hence a set-spectrum $F_Y = \sigma(T|Y)$ such that

$$\emptyset \neq F_Y \subset V(\lambda) \cap F^I.$$

We have

$$F_Y \subset F^I \subset F.$$

Since F_0 is the largest set-spectrum contained in F, $F_Y \subset F_0$ and hence

$$F_Y \subset V(\lambda) \cap F_0.$$

Thus, $\lambda \in \overline{F}_0 = F_0$, and inclusions (A.2.3) follow. □

A.2.4. *Corollary.* *Given* $T \in D(X)$, *if* F *is a closed subset of* $\sigma(T)$ *such that*

$$F = \overline{F^I},$$

then F *is a set-spectrum of* T. *In particular, every* $\overline{F^I}$ *is a set-spectrum of* T.

A.2.5. *Corollary.* *Given* $T \in D(X)$, *for every closed* $F \subset \sigma(T)$, *let* F_0 *be the largest set-spectrum contained in* F. *Then* $F-F_0$ *has void interior in* $\sigma(T)$.

A.2.6. *Corollary.* *Given* $T \in D(X)$, for every $F \in \mathcal{F}$,

$$F_1 = \overline{\sigma(T) - F}$$

is a set-spectrum of T.

A.2.7. *Corollary.* *Given* $T \in D(X)$, *let* $Y \in SM(T)$. *Then there exists* $Z \in SM(T)$ *such that*

$$\sigma(T|Z) = \sigma(T^Y). \tag{A.2.4}$$

Proof. By Proposition 11.10,

$$\sigma(T^Y) = \overline{\sigma(T) - \sigma(T|Y)}$$

and then Corollary A.2.6 implies that $\sigma(T^Y)$ is a set-spectrum of T. Then

$$Z = X_T[\sigma(T^Y)] ,$$

and (A.2.4) follows. □

A.2.8. Theorem. $T \in B(X)$ *is decomposable iff for every cover* $\{G_i\}_1^n$ *of* $\sigma(T)$ *with the* $G_i \subset \sigma(T)$ *open in the topology of* $\sigma(T)$, *there is a system* $\{Y_i\}_1^n \subset SM(T)$ *satisfying the following spectral decomposition*

$$X = \sum_{i=1}^{n} Y_i,$$

$$\sigma(T|Y_i) = \overline{G}_i, \quad i = 1,2,\ldots,n.$$

Proof. The "only if" part is obvious. Assume that $T \in D(X)$ and let $\{H_i\}_1^n$ be an open cover of $\sigma(T)$. There is a corresponding system $\{Z_i\}_1^n \subset SM(T)$ such that

$$X = \sum_{i=1}^{n} Z_i,$$

$$\sigma(T|Z_i) \subset H_i, \quad i = 1,2,\ldots,n.$$

For every i,

$$\overline{G}_i = \overline{H}_i \cap \sigma(T)$$

is a set-spectrum of T and $Y_i = X_T(\overline{G}_i) \in SM(T)$ with

$$\sigma(T|Y_i) = \overline{G}_i.$$

Moreover, the inclusions

$$\sigma(T|Z_i) \subset H_i \cap \sigma(T) \subset \overline{G}_i = \sigma(T|Y_i)$$

imply that

$$Z_i \subset Y_i, \quad i = 1,2,\ldots,n. \quad \square$$

In view of Proposition 3.6, any intersection of set-spectra is again a set-spectrum. For arbitrary unions of set-spectra we also have a corresponding closure property. To obtain it we need the following

A.2.9. Lemma. Given $T \in D(X)$, *for every subset* W *of* X,

$$F = \overline{\bigcup_{x \in W} \sigma(x,T)}$$

is a set-spectrum of T.

Proof. First, we show that $W \subset X_T(F)$. Indeed, for every $x \in W$, we have

$$x \in X_T[\sigma(x,T)] \subset X_T(F).$$

Next, with the help of Theorem 1.9 we obtain successively

$$\sigma[T|X_T(F)] = \bigcup_{x\varepsilon X_T(F)} \sigma(x,T) \supset \bigcup_{x\varepsilon W} \sigma(x,T) = F. \quad \square$$

A.2.10. *Theorem. Given* $T \varepsilon D(X)$, *let* $\{F_\alpha\}_{\alpha\varepsilon\Lambda}$ *be an arbitrary family of set-spectra for* T. *Then*

$$F = \overline{\bigcup_{\alpha\varepsilon\Lambda} F_\alpha}$$

is a set-spectrum of T.

Proof. For every $\alpha \varepsilon \Lambda$, there is a spectral maximal space Y_α of T with

$$F_\alpha = \sigma(T|Y_\alpha) = \bigcup_{x\varepsilon Y_\alpha} \sigma(x,T|Y_\alpha) = \bigcup_{x\varepsilon Y_\alpha} \sigma(x,T).$$

By denoting

$$W = \bigcup_{\alpha\varepsilon\Lambda} Y_\alpha$$

we obtain

$$F = \overline{\bigcup_{\alpha\varepsilon\Lambda} F_\alpha} = \overline{\bigcup_{\alpha\varepsilon\Lambda} \sigma(T|Y_\alpha)} = \overline{\bigcup_{x\varepsilon W} \sigma(x,T)}.$$

By Lemma A.2.9, F is a set-spectrum of T. $\square$

Part of this section is based on Erdelyi [2]. Theorems A.2.2 and A.2.3 with the Corollaries A.2.4 and A.2.5 are due to Berruyer [1]. Theorem A.2.8 and Lemma A.2.9 are among the results obtained by Bacalu [1].

A.3. *The approximate point spectrum and the single-valued extension property.*

Several examples as for instance the unilateral shift operators show that the SVEP does not imply the equality between the approximate point spectrum and the entire spectrum of an operator. A condition for such a property to hold is given in this section of the Appendix.

It is well-known that the approximate point spectrum contains the boundary of the entire spectrum. An extension of this property to the local spectrum now follows.

A.3.1. *Theorem. Let* T *have the* SVEP. *If* x *is any nonzero vector in* X *then*

$$\partial\sigma(x,T) \subset \sigma_a(T).$$

Proof. Let $\lambda_0 \in \partial\sigma(x,T)$. Then the local resolvent $\tilde{x}$ is unbounded near λ_0. In fact, if λ_0 is an isolated point of $\sigma(x,T)$ and $\tilde{x}$ were bounded in a neighborhood V of λ_0 then by

$$(\lambda-T)\tilde{x}(\lambda) = x, \quad \lambda \in V \cap \rho(x,T),$$

λ_0 would be a removable singularity of $\tilde{x}$. But then $\tilde{x}$ extends to λ_0 and this contradicts the maximality of $\tilde{x}$. If λ_0 is not an isolated singularity then it is an essential singularity of $\tilde{x}$ and again $\tilde{x}$ is unbounded near λ_0.

Thus we may choose a sequence $\{\lambda_n\} \subset \rho(x,T)$ such that $\lambda_n \to \lambda_0$ and $\|\tilde{x}(\lambda_n)\| \to \infty$. Then

$$y_n = \frac{\tilde{x}(\lambda_n)}{\|\tilde{x}(\lambda_n)\|}$$

is a sequence of unit vectors in X such that

$$(\lambda_n-T)y_n = \frac{x}{\|\tilde{x}(\lambda_n)\|} \to 0 \quad \text{when } n \to \infty.$$

Thus

$$(\lambda_0-T)y_n = (\lambda_0-\lambda_n)y_n + (\lambda_n-T)y_n$$

approaches zero as $n \to \infty$ and hence $\lambda_0 \in \sigma_a(T)$. □

A.3.2. *Definition. Let* T *have the* SVEP. *We say that the set-valued mapping* $\sigma : x \to \sigma(x,T)$, $x \in X$, *is localized if for every open* $G \subset C$ *with*

$$G \cap \sigma(T) \neq \emptyset$$

there is a nonzero $x \in X$ *such that*

$$\sigma(x,T) \subset G.$$

A.3.3. *Theorem. If* T *has the* SVEP *and* $\sigma : x \to \sigma(x,T)$ *is localized then*

$$\sigma_a(T) = \sigma(T).$$

Proof. Let $\lambda \in \sigma(T)$. By hypothesis, for every n, there is a nonzero vector $x_n \in X$ such that

$$\sigma(x_n,T) \subset D_n = \{\mu : |\mu - \lambda| < \frac{1}{n}\}.$$

Since the local resolvent is unbounded near any boundary point of the local spectrum, for every n we can choose $\lambda_n \in \rho(x_n,T)$ such that

$$d[\lambda_n, \sigma(x_n,T)] < \frac{1}{n}$$

and

$$\|\tilde{x}_n(\lambda_n)\| \to \infty \text{ when } n \to \infty.$$

Then for each n,

$$(\lambda_n - T)\tilde{x}_n(\lambda_n) = x_n$$

and we have unit vectors

$$y_n = \frac{\tilde{x}_n(\lambda_n)}{\|\tilde{x}_n(\lambda_n)\|}$$

such that

$$(\lambda - T)y_n = (\lambda - \lambda_n)y_n + (\lambda_n - T)y_n \to 0 \quad \text{as} \quad n \to \infty.$$

Hence $\lambda \in \sigma_a(T)$ and the proof is complete. □

For operators with the SDP the hypotheses of Theorem A.3.3 are satisfied. In fact, T with the SDP has the SVEP and then Lemma 4.4 implies that $\sigma : x \to \sigma(x,T)$ is localized.

A.4. *Some open problems.*

The importance of the SVEP in spectral decompositions has been seen throughout this work. The pathology lurking in operators which do not enjoy this property is brought out by certain "residual" parts of the spectrum which obstruct the construction of a satisfactory spectral theory. A thorough analysis of this case from the spectral decomposition standpoint was made by Vasilescu [2,3,4] with a follow-up by Bacalu [1].

Problem 1. Does the general asymptotic spectral decomposition imply the SVEP ?

The duality theory for decomposable operators has a most satisfactory conclusion expressed by Corollary 12.20. It seems that the class of quasidecomposable operators comes closest to the decomposable operators from the spectral decomposition standpoint. A hint that some kind of duality theory may exist for the even more general weakly and analytically decomposable operators is the fact that the

SVEP of the given T implies $\sigma(T^*) = \sigma_a(T^*)$, (Corollary 1.4).

Problem 2. What can be said about the dual of a quasidecomposable operator? Can any kind of duality theory be constructed for the class of quasi-decomposable operators?

The foregoing comment can be expressed for operators with the SDP giving rise to

Problem 3. How much can be said about a duality theory for operators with the SDP?

The axiomatic definition of the general spectral decomposition (Chapter II) was rewarding in properties and obviously there is much more to be said about it. One would expect that the decomposable operators defined in terms of the strongly specialized spectral maximal spaces form a proper subclass of the operators with the SDP.

Problem 4. Give an example of an operator with the SDP *which is not decomposable.*

As it was mentioned, all properties of operators considered in Chapter II hold for operators with the 2-SDP. The question arises whether there is any other property of operators with the SDP which does not hold for operators with the 2-SDP. In other words, we formulate our

Problem 5. Does the 2-SDP *imply the* SDP?

The spectral theory of weakly decomposable operators gained some remarkable properties by the additional decomposable spectrum hypothesis. The weakly decomposable operators do not seem to possess automatically decomposable spectra as the quasidecomposable operators do.

Problem 6. Is there a proper class of operators between the weakly- and the quasi-decomposable operators which possess the decomposable spectrum property?

The concept of spectral capacity in its original, weak and strong formulations characterizes a specific spectral decomposition.

Problem 7. Is there a suitable generalization of the spectral capacity adaptable to the general spectral decomposition (i.e. to operators with the SDP)?

BIBLIOGRAPHY

E. Albrecht

[1] An example of a weakly decomposable operator which is not decomposable. *Rev. Roum. Math. Pures Appl.* 20 (1975), 855-861.

E. Albrecht and F.-H. Vasilescu

[1] On spectral capacities. *Rev. Roum. Math. Pures Appl.* 19 (1974), 701-705.

C. Apostol

[1] Some properties of spectral maximal spaces and decomposable operators. *Rev. Roum. Math. Pures Appl.* 12 (1967), 607-610.

[2] Teorie spectrală și calcul funcțional. *Studii Cerc. Mat.* 20 (1968), 635-668.

[3] Restrictions and quotients of decomposable operators in a Banach space. *Rev. Roum. Math. Pures Appl.* 13 (1968), 147-150.

[4] Remarks on the perturbation and a topology for operators. *J. Funct. Anal.* 2 (1968), 395-409.

[5] Roots of decomposable operator-valued analytic functions. *Rev. Roum. Math. Pures Appl.* 13 (1968), 433-438.

[6] Spectral decompositions and functional calculus. *Rev. Roum. Math. Pures Appl.* 13 (1968), 1481-1528.

I. Bacalu

[1] Some properties of decomposable operators. *Rev. Roum. Math. Pures Appl.* 21 (1976), 177-194.

R.G. Bartle

[1] Spectral localization of operators in Banach spaces. *Math. Ann.* 153 (1964), 261-269.

[2] Spectral decomposition of operators in Banach spaces. *Proc. London Math. Soc.* (3) 20 (1970), 438-450.

R.G. Bartle and C.A. Kariotis

[1] Some localizations of the spectral mapping theorem. *Duke Math. J.* 40 (1973), 651-660.

J. Berruyer

[1] Opérateurs réguliers spectraux et opérateurs réguliers décomposables. *Publications du Département de Mathématiques, Faculté des Sciences de Lyon*, 1968, Tome 5, fasc. 2.

E. Bishop

[1] A duality theorem for an arbitrary operator. *Pacific J. Math.* 9 (1959), 379-397.

I. Colojoară

[1] Generalized spectral operators. *Rev. Roum. Math. Pures Appl.* 7 (1962), 459-465.

[2] Elemente de Teorie Spectrală. *Ed. Acad. Rep. Soc. Rom.*, București 1968.

I. Colojoară and C. Foiaș

[1] Quasi-nilpotent equivalence of not necessarily commuting operators. *J. Math. Mech.* 15 (1965), 521-540.

[2] The Riesz-Dunford functional calculus with decomposable operators. *Rev. Roum. Math. Pures Appl.* 12 (1967), 627-641.

[3] Theory of Generalized Spectral Operators. *Gordon & Breach*, New York 1968.

N. Dunford

[1] Spectral theory I. Convergence to projections. *Trans. Amer. Math. Soc.* 54 (1943), 185-217.

[2] Spectral theory II. Resolution of the identity. *Pacific J. Math.* 2 (1952), 559-614.

[3] Spectral operators. *Pacific J. Math.* 4 (1954), 321-354.

N. Dunford and J.T. Schwartz

[1] Linear Operators, Part I (1967), Part II (1967), Part III (1971), *Wiley* New York.

I. Erdelyi

[1] Unbounded operators with spectral capacities. *J. Math. Anal. Appl.* 52 (1975), 404-414.

[2] The set-spectra of decomposable operators. To appear.

[3] A class of weakly decomposable unbounded operators. *Atti Accad. Naz. Lincei Cl. Fis. Mat. Natur.* to appear.

I. Erdelyi and R. Lange

[1] Operators with spectral decomposition properties. *J. Math. Anal. Appl.* to appear.

I. Erdelyi and F.R. Miller

[1] Decomposition theorems for partial isometries. *J. Math. Anal. Appl.* 30 (1970), 665-679.

J.K. Finch

[1] The single valued extension property on a Banach space. *Pacific J. Math.* 58 (1975), 61-69.

S.R. Foguel

[1] The relations between a spectral operator and its scalar part. *Pacific J. Math.* 8 (1958), 51-65.

C. Foiaş

[1] Une application des distributions vectorielles a la théorie spectrale. *Bull. Sci. Math.* (2) 84 (1960), 147-158.

[2] Spectral maximal spaces and decomposable operators in Banach spaces. *Arch. Math.* (Basel) 14 (1963), 341-349.

[3] Spectral capacities and decomposable operators. *Rev. Roum. Math. Pures Appl.* 13 (1968), 1539-1545.

[4] On the maximal spectral spaces of a decomposable operator. *Rev. Roum. Math. Pures Appl.* 15 (1970), 1599-1606.

[5] On the scalar parts of a decomposable operator. *Rev. Roum. Math. Pures Appl.* 17 (1972), 1181-1198.

Ş. Frunză

[1] A duality theorem for decomposable operators. *Rev. Roum. Math. Pures Appl.* 16 (1971), 1055-1058.

[2] The single-valued extension property for coinduced operators. *Rev. Roum. Math. Pures Appl.* 18 (1973), 1061-1065.

[3] A new result of duality for spectral decompositions. *Ind. Univ. Math. J.* 26 (1977), 473-482.

[4] Spectral decomposition and duality. *Ill. J. Math.* to appear.

A.A. Jafarian

[1] Weak and quasi-decomposable operators. *Rev. Roum. Math. Pures Appl.* 22 (1977), 195-212.

A.A. Jafarian and F.-H. Vasilescu

[1] A characterization of 2-decomposable operators. *Rev. Roum. Math. Pures Appl.* 19 (1974), 657-663.

R. Lange

[1] Roots of almost decomposable operators. *J. Math. Anal. Appl.* 49 (1975), 721-724.

[2] Analytically invariant subspaces in spectral decompositions of linear operators on Banach spaces. Ph.D. Thesis, *Temple University*, Philadelphia 1974.

[3] 2-decomposable operators are decomposable. *J. Math. Anal. Appl.*, to appear.

[4] Duality theorems for decomposable operators of weak and strong type. Preprint.

F.-Y. Maeda

[1] Generalized spectral operators on locally convex spaces. *Pacific J. Math.* 13 (1963), 177-192.

S. Plafker

[1] On decomposable operators. *Proc. Amer. Math. Soc.* 24 (1970), 215-216.

M. Radjabalipour

[1] On decomposition of operators. *Mich. Math. J.* 21 (1974), 265-275.

[2] Equivalence of decomposable and 2-decomposable operators. Preprint. Abstract in *Notices Amer. Math. Soc.* 24 (1977), A-488.

J.E. Scroggs

[1] Invariant subspaces of a normal operator. *Duke Math. J.* 26 (1959), 95-111.

R.C. Sine

[1] Spectral decomposition of a class of operators. *Pacific J. Math.* 14 (1964), 333-352.

J.G. Stampfli

[1] Analytic extensions and spectral localization. *J. Math. Mech.* 16 (1966), 287-296.

[2] A local spectral theory for operators. V: Spectral subspaces for hyponormal operators. *Trans. Amer. Math. Soc.* 217 (1976), 285-296.

J.L. Taylor

[1] A joint spectrum for several commuting operators.*J. Funct. Anal.* 6 (1970), 172-191.

F.-H. Vasilescu

[1] On an asymptotic behaviour of operators. *Rev. Roum. Math. Pures Appl.* 12 (1967), 353-358.

[2] Residually decomposable operators in Banach spaces. *Tôhoku Math. J.* 21 (1969), 509-522.

[3] Residual properties for closed operators on Fréchet spaces. *Ill. J. Math.* 15 (1971), 377-386.

[4] On the residual decomposability in dual spaces.*Rev. Roum. Math. Pures Appl.* 10 (1971), 1573-1587.

AUTHOR INDEX

SUBJECT INDEX